The Unnoticed Challenge:
Soviet Maritime Strategy and the Global Choke Points

Robert J. Hanks

Special Report
August 1980

INSTITUTE FOR FOREIGN POLICY ANALYSIS, INC.
Cambridge, Massachusetts, and Washington, D.C.

Requests for copies of IFPA Special Reports should be addressed to the Circulation Manager, Special Reports, Institute for Foreign Policy Analysis, Central Plaza Building, Tenth Floor, 675 Massachusetts Avenue, Cambridge, Massachusetts 02139. (Telephone: 617-492-2116). Please send a check or money order for the correct amount along with your order.

Standing orders for all Special Reports will be accepted by the Circulation Manager. Standing order subscribers will automatically receive all future Special Reports as soon as they are published. Each Report will be accompanied by an invoice.

IFPA also maintains a **mailing list** of individuals and institutions who are notified periodically of new Institute publications. Those desiring to be placed on this list should write to the Circulation Manager, Special Reports, at the above address.

A list of IFPA publications appears on the inside back cover.

The Institute for Foreign Policy Analysis, Inc., incorporated in the Commonwealth of Massachusetts, is a tax-exempt organization under Section 501(c)(3) of the U.S. Internal Revenue Code, and has been granted status as a publicly-supported, nonprivate organization under Section 509(a)(1). Contributions to the Institute are tax-deductible.

Price: $6.50

Library of Congress No. 80-83751

ISBN 0-89549-025-0

First Printing
Printed by Corporate Press, Inc., Washington, D.C.

Contents

Summary Overview

In recent years it has gradually become evident to the American body politic that the United States is critically dependent on natural resources from overseas. The shock of the oil embargo, incident to the 1973 Arab-Israeli war, made the first real dent in what had, until then, been an altogether unjustified complacency pervading American public opinion. Even this traumatic experience, however, did not convince the bulk of the population that an energy crisis really existed. It took the better part of a decade, replete with recurring shortages in the availability of petroleum, to drive this basic lesson home.

More dimly perceived at present is an equally worrisome problem centered on other raw materials. Too few Americans realize that complete solution of the energy dilemma would ultimately be for naught if the nation's factories were forced to shut down because they could no longer obtain those metals and minerals without which the modern industrial process cannot function. Yet, insofar as a number of critical materials are concerned, the American dependence situation is nearly as unsettling as is the present petroleum outlook.

The foregoing factors lead inevitably to consideration of the U.S. position in this increasingly interdependent world and the importance of the globe's sea lanes to American economic and political survival. Assured access to sources of supply—as well as to markets for exports, the latter crucial to the nation's ability to pay for needed imports—lies at the heart of the problem, for a new threat to that assured access has arisen over the past two decades—a menace which has gone almost unremarked in the West in general and in the United States in particular. The threat is manifested primarily in the appearance of a large, modern, blue-water navy wearing the Hammer and Sickle of the Soviet Union. For the first time since the Bolshevik Revolution, Moscow now boasts a growing capability to sever those vital sea lanes.

This report examines the newly developing threat to the future well-being of an insular country such as the United States—despite its continental size, nonetheless an island nation—posed by a land-bound, predatory country which has only recently discovered the importance of sea power and the world's sea lanes. It does so by (1) tracing the rise of Soviet maritime power, (2) recalling the historical evolution of sea-going nations, leading to the ultimate power established by the British Empire, (3) outlining the evolution of Soviet maritime thought, (4) setting forth the pattern of Soviet politico-maritime advances in recent years and describing the implications of that Soviet campaign, (5) comparing U.S. and Soviet naval forces, (6) analyzing the American "swing strategy" as it presently exists and what it means for the global maritime position of the United States as well as that of the North Atlantic Alliance, and (7) discussing measures which should be taken to meet this little noticed challenge on the high seas.

The Rise of Soviet Sea Power

Although there has been a lively debate in the United States over the question of just when the Soviet naval buildup began, most informed observers place its origins in the late 1950s, following Nikita Khrushchev's appointment of Admiral of the Fleet of the Soviet Union Sergei G. Gorshkov to head the Red Navy. Under orders from Khrushchev to scrap the surface fleet and concentrate solely on rocket-firing submarines, Gorshkov—today in his third decade as commander of the Soviet Navy—did indeed build the submarine fleet. Not only outlasting his mentor but outmaneuvering him as well, Gorshkov also built a powerful new surface fleet, one that today outnumbers the U.S. Navy by an order of magnitude. Moreover, he has made steady progress in the past ten years in developing the two arms of naval power which heretofore have symbolized clear and unchallenged U.S. superiority over other navies, especially that of the USSR: aircraft carriers and amphibious forces. Of course, along with the remarkable growth in Soviet naval prowess, Moscow has made equally significant strides with respect to the other components of maritime power: the merchant marine, oceanographic capability, and a huge fishing fleet.

If one now couples these advances with an examination of the changing deployment patterns of the Soviet Navy, a disconcerting picture emerges, one suggesting that Moscow is well along the road to achieving a substantial degree of maritime hegemony on a global basis—a situation which bodes ill for an insular nation such as the United States.

Patterns of Maritime Dominance

If one looks backward into history, a good deal of illumination can be shed on the current maritime actions of the Soviet Union. Commencing with the great ages of exploration and the pioneering work of the Portuguese visionary, Prince Henry the Navigator, man began to comprehend the nature and extent of the globe which he inhabits. As Henry's captains pushed the maritime frontiers ever outward, the realities of wind and wave together with certain geographic truths gradually defined the tracks which ocean-going ships—merchant as well as naval—should follow during their myriad voyages. As sea-going experience accumulated, the notion of shipping choke points entered the ken of man. Sometimes formed by proximate land masses—as in the case of the Strait of Gibraltar—and sometimes by a combination of weather and a land mass—as in the case of the Cape of Good Hope—these choke points developed into maritime funnels through which the great bulk of all shipping passed.

No nation understood the significance of these choke points better than did England. As the Empire expanded, the British Government, relying primarily on the ships of the Royal Navy, reached for control of these choke points. At the height of England's colonial power, domination of those constricted passages had become an established fact. One can recall old geography books in which the British Empire was invariably portrayed in pink colors, and those maps disclosed that all the major choke points were bracketed or otherwise dominated by that omnipresent color.

It makes little difference, of course, whether the leaders of the Soviet Union have been studying the history of the British Empire or reading the works of the American naval prophet—Alfred Thayer Mahan. What is significant is that the pattern of Soviet politico-maritime initiatives in recent years suggests very strongly that Moscow has taken aboard the lessons derivable from such study. Thus, an examination of the world's choke points today finds the Soviet Union busily extending its influence—indeed, in some cases, dominion—over many of them.

The Evolution of Soviet Maritime Thought

The roots of the Soviet Navy can be traced to the role it played during the Bolshevik Revolution—essentially, infantry ashore. Aside from manning a small riverine force, Soviet sailors fought mainly as land-bound soldiers. Moreover, since the new governmental leadership was drawn exclusively from Red Army units, the Soviet Navy was viewed as a force useful solely for protection of the salt-water flanks of that Army. This position in the military hierarchy was to be sustained until the advent of Gorshkov.

Gorshkov was immeasurably aided in his ambitions by Stalin's experience in the wake of the Second World War. Although the Soviet Union emerged from that conflict as one of the two most powerful nations on earth, the dictator in the Kremlin nonetheless found himself invariably frustrated by salt water whenever and wherever he sought to extend

Soviet influence beyond the Eurasian land mass. Despite the fact that Khrushchev was also a product of the Red Army, Gorshkov must have hammered home the lesson he had learned over the years and which the remainder of the Kremlin leadership at last came to accept: Wherever the Soviet Army reached the shores of Mackinder's "heartland" it ran afoul of salt water. Given the subsequent expansion of the Soviet Navy, one can only assume that Gorshkov expounded on this theme at every opportunity. In any event, the world is today witness to the development of a true blue-water capability on the part of the Soviet Union. How ironic that such a nation—heretofore captive of an essentially land-bound mentality—has at last come to the realization which maritime countries have grasped for centuries: Any nation which aspires to global dominance must learn to exploit the seas.

The Pattern of Soviet Advances

A survey of global choke points, commencing in the Caribbean Sea and progressing around the world, reveals that, with respect to most of the shipping constrictions once controlled by the British Empire, the Soviets can now be found working assiduously to establish their influence. Whether in Cuba—at the crossroads of the Caribbean—or in the Indian Ocean at Aden, as well as at many other places, the Soviet politico-maritime hand is in evidence. Thus, coupling the growth of the Soviet Navy and the other facets of Moscow's maritime power with the implications of the extension of Soviet influence to regions containing historic shipping choke points, one can reasonably conclude that a concerted drive for maritime hegemony is in progress.

U.S. and Soviet Naval Comparison

The foregoing findings immediately raise questions with respect to the sea power balance existing between the United States and the Soviet Union. An analysis reveals that while Soviet naval and other maritime strength has steadily increased over the past two decades, that of the United States has undergone a rapid decline. Today the Soviet Navy holds a decided advantage over that of the United States in overall numbers and is rapidly closing extant gaps in aircraft carriers as well as in amphibious forces. There is now little doubt that, following the evolutionary development of the *Moskva* and *Kiev* class carriers, the USSR is currently constructing its first nuclear-powered, attack carrier which will approach 80,000 tons displacement. The follow-on construction of a number of these ships will go far toward erasing the most significant advantage the American Navy has possessed to the present.

Furthermore, while true cruiser construction in the United States ended some time ago, recent reports from the Soviet Union suggest that at least four new classes of cruisers are under construction, including a monstrous 30,000-ton strike cruiser, a smaller version of which was planned, but never built, by the United States.

It is true that the overall number of ships in the Soviet Navy has now begun to decline slightly and that Gorshkov seems to be concentrating on quality as opposed to quantity. Thus, it appears that future trends will produce a slightly smaller Soviet fleet—one not only still numerically superior to that of the United States, but with vastly increased sophistication and capability.

Meanwhile, despite recent statements to the contrary by Carter Administration officials, the outlook for a resurgence of American warship construction is decidedly unfavorable. What additional funding is being programmed for the armed forces is being weighted heavily toward the U.S. commitment to the defense of NATO's Central Front rather than to meeting the challenges the United States and the Alliance confront at sea, not only in the North Atlantic but elsewhere around the globe.

The Swing Strategy

The latter factor—concentration of U.S. attention almost exclusively on NATO's Central Front—is most starkly manifested in the so-called American "swing strategy." Shortage of armed forces has led the U.S. Government, over the years, to adopt a strategy wherein any war in Europe between NATO and the Warsaw Pact could instantly trigger a sizable shift of American military units—primarily naval—from the Pacific to the Atlantic theater.

The commander of all U.S. military forces in the Pacific Ocean area has stated that his presently assigned strength is only marginally capable of meeting those commitments for which he is responsible and that, if significant segments were to be withdrawn for service in the Atlantic, he would have no option but to adopt a defensive posture—abandoning the Western Pacific—in order to survive. In response to their civilian superiors in the Department of Defense, the Joint Chiefs of Staff have let it be known that, in their collective opinion, the answer is to build forces—not "swing" them.

The impact such a U.S. abandonment of the Western Pacific would have is not difficult to envision. After all, Japan is governed by a constitution—essentially imposed by the United States at the end of the Second World War—which forbids it the military prowess necessary to defend itself unilaterally against foreign aggression. Since 1945, the Japanese have necessarily relied on military protection provided by the United States. Removal of this defensive umbrella would surely cause grave security problems for that island nation. In the event of a European war—and there are many who argue that such a conflict would inevitably become global in nature—an American military retreat from the Western Pacific would likely cause Tokyo to assume a neutral stance, at best, or to seek some sort of accommodation with the Soviet Union, at worst. The effect on the People's Republic of China—a neutral or Soviet-dominated Japan on one flank and a major enemy on the other—is equally if not more worrisome. It is not too difficult, for instance, to imagine Peking pursuing some sort of rapprochement with the USSR, an action that would profoundly alter the worldwide balance of power. One has only to recall the reality that over 40 Soviet divisions and a commensurate number of aircraft squadrons are currently tied down in the Pacific Far East by Moscow's perception of the Chinese threat, forces that otherwise could be quickly shifted to the European front. The increased dangers which would then confront NATO are manifest. Similar effects throughout the remainder of the Far East and the all-important Indian Ocean make for a chilling prospect.

Meeting the Challenge

Two questions move center stage. First, what is the Carter Administration doing about the problem? Second, what ought to be done?

With regard to the Carter Administration, the policy clearly is one of retrenchment. The naval shipbuilding program left by President Ford has been cut back appreciably by the new Administration. Moreover, Jimmy Carter consistently vetoed any proposal to build additional aircraft carriers, he and his advisers claiming that this class of ship was far too expensive and far too vulnerable in the modern naval environment. Yet, it is interesting to note that these ships have proved to be the first upon which President Carter has called whenever a crisis has erupted. Subsequent naval shipbuilding programs devised by his advisers have served merely to confirm the thrust of the initial Carter budget. Despite subsequent crises in Iran and Afghanistan, the pattern persists.

What ought to be done? The first move should be to reconstitute the power of the United States Navy and other elements of American sea power. The dangerous decline which has been allowed—indeed, encouraged—in recent years must be reversed. The Soviet drive for domination of the world's sea lanes and choke points admits of no other rational course.

Next, the whole strength of the nation—political, economic and military—must be mobilized. In a power-political world, a single element of a nation's prowess is simply not enough to safeguard its global interests. To achieve success, the effort must call on the total force a nation disposes and that power must be integrated and applied in a carefully orchestrated manner. It makes little sense, for example, to spend huge sums on expansion of American military strength while, at the same time, the nation is exporting advanced technology to an adversary or failing to link that opponent's conduct elsewhere in the world to the basic threats posed to U.S. security.

It would be the ultimate irony of history if the United States—a nation with a 300-year heritage of dependence on and experience in the use of the seas, a nation whose dependence on the seas is greater than ever today—were ultimately to be bested by an international predator whose fundamental attention has, until recent years, been centered exclusively on the land.

1. Introduction

Inundated as they have been in recent years by numerous domestic problems and frequent international crises, most Americans have failed to notice the emergence of an insidious pattern of maritime-oriented developments—on a global scale—which, if unchecked, could one day gravely endanger, or perhaps lead to the demise of, the United States of America as a free and democratic country. The supreme irony of this evolving situation lies in the fact that for more than two centuries the United States has been, first and foremost, an insular nation, heavily dependent on two-way mercantile trade—one enjoying a long and illustrious maritime heritage.

Since those long-ago days when the first European settlers struggled in wilderness outposts in Virginia and Massachusetts, peoples of this vast land have been tied to the sea. Almost totally dependent upon shiploads of tools and supplies from England, the early colonists necessarily looked seaward for survival. Over the ensuing decades, that dependence endured. Eventually contributing to insurrection and independence, it proved to be an important factor in spawning one of man's great experiments in democratic government.

For a brief period of time, as American attention turned west and the young nation inevitably expanded toward the second ocean which bounded the continent, many lost sight of the oceanic thralldom to which they were subject. But with their arrival on those distant Pacific shores, dependence on the seas emerged once more as a dominant facet of American life. The great clipper ships—solely a product of the innovative American mind and the nation's sea-going traditions—captured universal admiration by making incredible wind-driven voyages around Cape Horn to California and onward to China and Japan where the resources and markets of the fabled Orient lay waiting for exploitation.

Following the advent of the industrial revolution, it quickly became apparent that as rich in natural resources as this young nation was, there still were numerous commodities around the globe without which the newly developing American industries could not function. As the decades passed and domestic supplies of raw materials were progressively depleted, American dependence on overseas sources steadily increased. Moreover, exports remained the only feasible means of paying for those commodities. Fundamentally, conditions had changed little since the days of the colonists, and dependence on the sea thus continued. So it is today.

It is also ironic that the United States, which from its inception has been maritime-oriented, now finds its future clouded by a surreptitious challenge on the world's oceans, a challenge mounted by another nation which traditionally has been single-mindedly land-bound in its thinking and in its actions alike. For the challenge which increasingly threatens America's well-being is being posed by the Soviet Union.

As a result of the "squeaky-wheel" syndrome, one suspects, U.S. domestic attention has once again been drawn away from the inescapable fact of oceanic dependence. In the present instance, the foci have been inflation, unemployment, pollution and similar difficulties. Internationally, the American body politic has worried about those matters routinely capturing the current media headlines: danger of a nuclear war, fear of another Vietnam-type involvement, the decline of the dollar, the revolution in Iran, and the brutal Soviet invasion of Afghanistan. Even in the latter case, where the developing threat is to continued American access to Persian Gulf oil, there are many in this land who vociferously oppose any U.S. action in the region, claiming that vital national interests are not involved. As a matter of fact, one Democratic candidate for the presidency—Senator Edward M. Kennedy—stated that he is sure "every American would prefer to sacrifice a *little gasoline* (emphasis provided) rather than shedding American blood to defend OPEC pipelines in the Middle East."

Unnoticed in the face of this surfeit of problems, another threat has developed which, in the long term, could be of far greater consequence than most of the difficulties presently occupying center stage: The United States is in serious danger of losing its historic ability to freely use the sea lanes which crisscross the planet. To an insular nation—as the United States indisputably has always been and remains—herein lies a prescription for disaster. If the citizens of this country ever lose their access to raw materials from overseas, to markets for their products, or to allies beyond the oceans, the end of the great American experiment will not likely be far behind.

2. The Rise of Soviet Sea Power

After several decades of concentrating on massive ground and air forces, while relegating its navy to the role of protecting the seaward flanks of the Red Army, the Soviet Union began to develop a blue-water naval fleet. As the latter capability has expanded, the Russian Bear has begun to reach for control of the world's oceans, something heretofore deemed to be the exclusive province of major maritime powers—primarily Western—such as Great Britain and the United States.

The remarkable rise which the Soviet Navy has experienced in recent years has been engineered primarily by Admiral of the Fleet of the Soviet Union Sergei G. Gorshkov, who has commanded the Red Navy since 1956. Quite obviously, Gorshkov succeeded in convincing the hierarchy in the Kremlin that the USSR could not pretend to global power status without the means to control the seas—Alfred Thayer Mahan's "broad common" across which travels most of the planet's commerce and military prowess.

One can find considerable disagreement with respect to the period in which the present-day Soviet naval transformation began. Many will argue that the Cuban missile crisis of 1962 marked the great watershed in Soviet naval policy. This group contends that the humiliating events of those tense days dramatically demonstrated to the Soviet leadership the indispensability of viable sea power; that from that time onward, Moscow—impelled by that traumatic experience—set out under forced draft to acquire this kind of maritime strength.

Others contend, with considerable persuasiveness, that the change in Soviet naval thinking occurred somewhat earlier. The theory espoused by the latter group is founded on the amount of time generally agreed necessary to carry new warship designs from the conceptual to the hardware stage; that is to say, from first design drawings to a lead ship, completed, afloat and operating. Pointing to the appearance of the first modern, guided-missile surface ships in the Soviet Navy in the 1962-1963 time period—*Kynda*-class cruisers and *Kashin*-class destroyers—then working backward through the build, design and concept stages, this group places the Soviet naval expansion origins sometime around the mid-1950s. This would have been shortly after Gorshkov's appointment to head the Soviet Navy. Moreover, it would have been in the wake of the 1956 Suez crisis which highlighted Moscow's inability to intervene in the face of American sea power in the Mediterranean.

Available evidence suggests that the latter group's view is much nearer to the truth. On the other hand, one can be reasonably certain that the Cuban missile crisis added decided impetus to the Russian naval expansion program—if any were still needed.

A second dispute centers on the mission of the emergent Soviet Navy. One view holds that the recent Soviet naval expansion originated with and still is driven by the American sea-based strategic threat to the USSR. Citing the anti-submarine classification which the Kremlin has continued to apply to all major, new-construction warships—including the aircraft carrier *Kiev*—theorists taking this position contend that Moscow remains wedded to its traditional naval strategy: defense of the Russian homeland. The notion is that neutralization of the U.S. sea-launched ballistic missile threat, posed by American Polaris and Poseidon submarines, is the prime mission of the entire Soviet Navy. Such argumentation is usually buttressed by pointing to the Soviet interest in countering the American aircraft carriers when these ships bore the full load of the sea-borne strategic mission, and then to the Soviet change of emphasis to anti-submarine warfare as the threat shifted to the submarine-launched ballistic missile.

This view cannot be dismissed out of hand and, to be sure, might ultimately prove to be correct. Nevertheless, there is a substantial body of contrary evidence suggesting that Soviet leaders have come to the broader realization that sea power offers a wide range of opportunities for the promotion of their national interests and objectives around the globe. Moreover, it appears that Moscow has embarked on a deliberate campaign to exploit those opportunities.

While the stunning growth of the Soviet Navy in size and sophistication—under Gorshkov's tutelage—has caught the attention of the world's navalists, and certainly must be recognized as a major factor in the international equation, that expansion does not constitute the sole cause for concern. In terms of global maritime control, the mosaic is far more intricate. For the Soviet reach for the oceans entails diplomacy, intrigue, economics, ideology and national determination, in addition to acquisition of military as well as naval and other forms of maritime hardware. Moreover, the drive clearly is worldwide in scope.

One cannot, of course, view sea power solely in naval terms. As Alfred Thayer Mahan clearly recognized, far more is involved than a seventeenth century ship-of-the-line or its modern counterpart, the nuclear-powered aircraft carrier. A broad definition of sea power encompasses not only warships but many other varieties of sea-going craft, including those mundane ships transporting cargoes of various kinds, those seeking the protein-laden life dwelling in the sea, or those craft probing the depths in an attempt to understand the anatomy of the oceans themselves. All of the foregoing categories can today be found in quantity in the Soviet sea power inventory. Moreover, the ships involved are numerous, modern and highly capable. To these Moscow has added yet another category: the sea-going, electronic ferret. No U.S. naval officer who has spent any time at sea in recent decades is unfamiliar with the ubiquitous Soviet "trawler"—an electronic intelligence collector disguised as an otherwise innocent fishing vessel—which almost continuously tracks U.S. naval forces.

As an aside, it is also worth noting that few Americans are aware of the fact that most long-distance telephone calls along the east and west coasts of the United States—those passing via micro-wave radio relay stations—are routinely monitored and recorded by these Soviet electronic snoopers. Even fewer Americans realize that their own naval ships are forbidden to trail the Soviet eavesdroppers and monitor the same frequencies to determine what the Soviets are learning about us—because to do so would constitute a domestic invasion of privacy. In short, the USSR is allowed to invade that privacy, but our government is prevented from assessing the damage associated with such foreign intrusions.

The foregoing survey confirms the fact of Soviet prowess in maritime fields other than naval. Yet, perhaps the facet of Soviet sea power least remarked in the United States is MorFlot—the state-owned and -operated merchant marine.

Here one encounters diametrically opposed U.S. and Soviet approaches to the conduct of international oceanic shipping. Historically, the United States has viewed its own merchant service as a small, peacetime industry susceptible to rapid expansion in times of emergency or war. Today, for example, no more than 6 percent of American overseas trade is carried in U.S.-flag ships. Moreover, the active American merchant fleet, comprising 517 ships and 13 million deadweight tons in 1975 (of which roughly 40 percent of the ships and 60 percent of the tonnage were devoted to the tanker trade) ranked tenth among the world's nations.

The Kremlin, on the other hand, has opted for the construction and maintenance of a large, modern fleet under total state control, operating continuously in direct support of Soviet national aims and objectives. Swiftly responsive to directions from Moscow, MorFlot ships can be manipulated to undercut international rate structures, diverted to support current naval operations, or otherwise utilized to further Soviet interests. At something in the neighborhood of 1,700 ships and 16,000,000 deadweight tons, MorFlot ranks only behind Japan and Liberia. The contrast in approaches could not be more evident—even without considering the various other facets of maritime power.

For instance, the Soviet Union presently operates the world's largest fishing fleet: more than 4,000 ocean-going vessels. Deploying groups of several hundred trawlers at a time, accompanied by modern factory ships, the USSR today fishes in all of the world's oceans, the annual catch being second only to that of Japan and roughly three times that of the United States.

In the field of oceanic research and surveying, the Soviet Union possesses in excess of 200 ships—more than the rest of the world combined. Wherever the present-day seaman goes, he can find these ships busily probing the depths and cataloging the information discovered.

Whether engaged in cut-rate incursions into the international maritime trade market, conducting gratis fishing resources surveys for some target country, or observing American missile tests, this vast array of normally non-naval ships is continuously operated directly in furtherance of Soviet state policy. And that policy frequently dictates maritime construction choices bearing little relation to the market demands which govern U.S. and other Western commercial decisions. For example, Moscow has elected to continue acquiring significant numbers of small, break-bulk cargo ships rather than concentrating exclusively on the much larger and more efficient container ship. The reason for this decision is relatively straightforward: the former are more suitable for operating in the small, primitive ports of many less developed nations around the world, especially those of the Third World countries the USSR seeks to exploit and subvert.

Given the balance manifested by the sea power trident currently wielded by the USSR, it seems clear that the fundamental tenets of this incomparable power have long since begun to be appreciated within the walls of the Kremlin. When the growth of the Soviet Navy is added to the foregoing picture, it becomes evident that Moscow has at last abandoned the exclusively land-bound mentality which ruled for so long in that nation.

In light of these developments, one can argue with considerable conviction that the Soviet leadership, most likely again prodded by Gorshkov, has learned a prime lesson from the history of the British Empire, and that the experience of the Suez and Cuban confrontations merely served to provide added incentive for Moscow to press ahead full speed in putting that lesson to practical use. If this notion is valid, its recognition and comprehension should be of extreme importance to those in whose hands rests the future security of the United States and, concommitantly, that of the remainder of the Free World. Those who currently debate the more readily apparent or more immediate U.S. national security issues would do well to devote some attention to the serious challenge represented by the increasingly obvious Soviet reach for international maritime hegemony.

While evolving American foreign and military policies seem to be designed to pull back United States commitments—except for those to defend Alaska, Hawaii and NATO's Central Front—to America's continental shores, the USSR has continued to expand Soviet maritime prowess in all its myriad aspects.

The Soviet invasion of Afghanistan—launched in December 1979—shocked the world in general and the American President in particular into the realization that we do, indeed, live in a power-political rather than an idealistic world. Goaded by this event, together with the earlier seizure of the U.S. Embassy in Teheran, Carter came reluctantly to the conclusion that the Soviet Union was pursuing a plan—not necessarily well conceived or executed with precision—to achieve global hegemony.

To understand the portents of any major shift in the balance of global maritime power, one must again glance into the rear-view mirror of history. If, as suggested in this study, Moscow has indeed taken a page from the book of Imperial England, an examination of that great historical era is indispensable to any comprehension of current Soviet aims and objectives. Such an examination must begin with the early years of the fifteenth century and the origins of the age of exploration.

3. Exploration and Empire

It was in the early years of the fifteenth century—at Sagres in Portugal, perched high above the windswept southwestern tip of Europe—that the recluse son of a Portuguese king sparked the first great age of discovery. The modern world owes a huge debt of gratitude to Prince Henry—known to history as The Navigator. Possessing a keen mind and a driving ambition, Prince Henry was also a man of vast vision. Seeking to circumvent the barrier between Europe and Asia erected by the legions of Islam, Henry turned seaward for an alternative route.

Gathering mathematicians, cartographers, sea captains, astronomers, and others around him at Sagres, he set to work improving navigation methods, charts, and the instruments by which his captains would find their way around Africa and into the Indian Ocean. To complete the material needs of the seamen he planned to send off into the unknown seas, a far more seaworthy ship was also needed. Even here the Prince's ingenious hand was evident, in the development and building of the famous Portuguese caravel which would eventually open the sea-route to India.

Prince Henry never sailed in the ships he sent forth, nor did he live to see the fulfillment of his dreams. But the work he and his entourage did on this bleak coast still stands as a monument to exploration. Urging his captains ever southward along the African coast, he ordered them back again and again. By the time of Henry's death in 1460, those ships had reached only a third of the way to the southernmost African cape—still thousands of miles away. But the impetus of four decades of Henry's urgings drove his Portuguese captains onward after his death, until Bartholomeu Dias doubled Cape Agulhas in 1487 and, at last, the maritime route to India lay open to European seamen.

Over the next century or so, led by the intrepid Portuguese sailors, exploratory activity intensified. Other nations, eager to learn about the Portuguese efforts and discoveries, flooded that country with spies as they sought to capitalize on the pioneering work of Prince Henry and his captains. Off sailed their own ships in search of the riches of the Indies and China, some by way of the newly discovered Cape of Good Hope, while yet others headed directly westward hoping to find what was believed by some to be a shorter route. By the time this initial period drew to a close, Columbus had reached America, Vasco da Gama had established trade with India, and Magellan had circumnavigated the globe.

Another burst of exploration, second only to the Columbus-Magellan era, erupted in the mid-eighteenth century and culminated in the three incredible voyages of Captain James Cook. The intervening decades, to be sure, witnessed continuing activity at varying levels and the list of sea-pioneers who combed the planet's oceans is a long one.

Individual motives for sailing off into those terror-shrouded, unknown seas may have been disparate and mixed, but national aims were certainly less confused: those who financed the expeditions expected—and demanded—a return on their investments. Even the visionary Prince Henry anticipated that the ultimate results of his efforts would promote Portuguese power and wealth. And as he also anticipated, increased knowledge would be an inevitable dividend.

From these voyages, global concepts emerged for the first time. Man at last began to perceive and then to comprehend the true shape as well as the vast extent of the celestial body he inhabited. Moreover, since trade and gain provided the fundamental drive behind the exploratory effort, establishment of the safest and shortest sea-routes became a primary goal. As knowledge continued to expand, the notion of maritime choke points—restricted passages through which most of the sea-borne traffic would eventually funnel—entered man's ken. Dias and his contemporaries, for example, quickly learned that they would do well to give a wide berth to the unpredictable currents off Cape Agulhas. In doing so, however, they could not stray too far south or they would encounter the heavy weather and mountainous seas of what came to be known as "Roaring Forties"—the high latitudes which produced winds sweeping around the globe across an essentially unbroken stretch of ocean. By trial and error, those pioneering captains had discovered that maritime choke points were not confined to the English Channel or the Strait of Gibraltar, nor were they always bounded on both sides by land masses. These mariners would find others as they made their way into and across the Indian Ocean.

Initially, interest in these choke points centered on the greater certainty provided by coastal navigation, shorter distances, and the increased safety inherent in protected waters as opposed to myriad hazards encountered in open, storm-swept seas. Soon thereafter, however, owners and sailors also began to worry about man-made dangers, with pirates and wreckers presenting the more common threats. This was especially true in and around narrow straits and off rocky capes. Eventually galvanized by anguished cries from their merchant princes, European nations began sending out warships to protect their relatively lightly armed and lumbering merchantmen against many of the man-made perils. Sheer necessity thus dictated that the flag would follow trade.

During the four centuries following Prince Henry, the first global powers appeared on the world's stage. They vied continuously with one another to claim and then, more importantly, to exploit the riches of the distant lands their courageous voyagers discovered. All of these countries were successful to one degree or another, of course, but one eventually surpassed all the rest: England. While the fortunes of Spain, Portugal, France and Holland dominated from time to time, those of the British Empire steadily prospered—with the notable exception of the American Revolution—from the days of Queen Elizabeth and Sir Francis Drake until the end of World War I. The heyday of English colonialism was reached during the reign of Queen Victoria when the sun, indeed, never set on the British Empire.

One feature of that empire is all too often overlooked: At the height of Great Britain's maritime power, the Royal Navy and England's vast colonial holdings stood astride every major shipping choke point on the face of the earth. One has only to recall the geography books of days long gone, when that particular empire was invariably portrayed in pink or red. A close examination of those maps reveals that all of the strategic maritime focii were bracketed by those colors: Gibraltar, Malta, Suez, the Cape of Good Hope, the Falkland Islands, Aden, Ceylon, Singapore, Hong Kong, Australia, islands throughout the Pacific and Caribbean—the list seems endless. Those maps indicate that quite early London had taken aboard an invaluable lesson in international power politics: dominance over shipping choke points confers two distinct advantages on their controllers. First, it safeguards the hegemonic nation's sea-borne traffic against depredations by others during transit through restricted waters where that traffic is most vulnerable to man-made threats. Far more important, that control carries with it the ability to regulate the traffic of every other nation using such choke points. No nation understood this truth better than Great Britain.

Lest anyone suffer from the delusion that the days when nations struggled for control over choke points ended with the demise of nineteenth century imperialism, the downfall of the British Empire, or the passing of Alfred Thayer Mahan, he ought to consider the following present-day developments: the American campaign to secure Western Hemispheric guarantees of unimpeded access to the Panama Canal; threats from Singapore, Malaysia and Indonesia to constrain oil-tanker traffic through the Strait of Malacca; Iran's seizure of Abu Musa and the Tunbs Islands for the advertised purpose of ensuring free passage through the Strait of Hormuz; and the fight which has raged in recent years in the United Nations over a new, comprehensive law of the sea. One of the most bitterly contested issues encompassed by the latter initiative has been the proposed expansion of territorial seas to the point that numerous shipping choke points—historically considered to be high seas and, therefore, open to free passage—would fall under the jurisdiction of the riparian states and, thereby, be subject to closure.

At the end of the age of discovery and following exploitation of the new lands, this unyielding verity remained: He who controls the world's maritime choke points wields incredible power—economic, political and military—over the destinies of others. As the nineteenth century came to a close, England clearly held that sort of power.

4. The Evolution of Soviet Maritime Thought

The Age of Exploration and Empire forms an illuminating prelude to an examination of Soviet politico-maritime initiatives in recent years. The emerging pattern of these moves first struck me during the early years of the 1970s when I commanded the U.S. Middle East Force, a small group of warships operating from the Persian Gulf shaikhdom of Bahrain. For two-and-a-half years, I lived in and traveled throughout the western half of the Indian Ocean, including the Gulf and the Red Sea. Part of my task was to keep track of the ships of the Soviet Navy and try to divine what they were up to. It did not take long for those old high school maps to come back to mind. It seemed that everywhere the Russians could currently be found active—politically and militarily, overtly and covertly—those maps had been colored pink or red. Had Admiral Sergei Gorshkov and the Soviet leadership taken a page from the book of Imperial England? Had the men in the Kremlin been studying the works of Mahan? At least in the Indian Ocean area, it appeared so. A glance at other portions of the globe seemed to confirm the notion.

While not so evident then as it is today, the pattern of Soviet politico-maritime incursions around the rim of the Indian Ocean could nonetheless be discerned in the early 1970s. A basic question thus materialized: What in the maritime world were the Soviets trying to do?

There are those, as noted earlier, who argue today that—Soviet strategic submarines aside—the Kremlin has built a Navy designed solely for anti-submarine warfare: one with a dual mission of defending the motherland against the manifest threat from American ballistic-missile submarines, and of protecting their Soviet counterparts against U.S. attack. Some others would add to that wartime role a peacetime mission: presence. The latter task, of course, constitutes useful political employment for the Soviet fleet pending the outbreak of a shooting war with the West. Still others maintain that the USSR is continuing to pursue a course leading to world dominance, and that Moscow—under two decades of nudging by Admiral Gorshkov—now recognizes the vital importance of the seas and sea power to that global objective. This notion has not, to be sure, always been either accepted or understood in the Kremlin.

Following the triumph of the Bolshevik Revolution, the most pressing matter for the leaders of the Soviet Communist Party was consolidation of that upheaval's gains. Their attention was necessarily focused inward, with internal control and stability receiving first priority. While concentrating on assuring the permanence of their internal revolution, Lenin and his lieutenants nevertheless kept a wary eye on external threats, especially since in the early days of the Revolution the Western allies of the Czar had intervened on Russian soil to support those factions opposed to the Bolsheviks.

Given the nature of the Russian Revolution and the great territorial expanse of the embryonic Soviet state, it is little wonder that the new leadership came to be dominated by the hierarchy of the Red Army. In light of this development and the nation's then relative self-sufficiency with respect to raw materials, it should come as no surprise that Bolshevik Russia evolved as an almost exclusively land-oriented power.

In geostrategic terms, it would appear that Lenin read and embraced Sir Halford J. Mackinder's contemporary *Democratic Ideals and Reality*. Certainly, he believed, as Mackinder did, that the world island was the key to world dominion and that its heartland could not be outflanked by sea power. Such a land-oriented mind-set would view naval strength as little more than an adjunct to the more relevant power of armies and the then emerging air forces. The Soviet Navy's sole mission could therefore be expected to encompass little more than guarding the heartland's seaward flanks against depredations perpetrated by maritime-oriented enemy states. With that notion reinforced by the experience of the Revolution, when the Czar's recent allies intervened around the periphery of the infant union of Soviet states, Lenin relegated the Soviet Navy to this precise role. Thus, the Red Navy, whose sailors had ultimately wound up fighting ashore as infantry during the Revolution, was henceforth quite naturally looked upon by the new power structure as good only for providing protection to the salt-water flanks of the Red Army.

Possessing no real sea-going traditions except those deriving from the days of the Czars—no good communist would ever lean on such a crutch—the Soviet Navy had to grow and mature under the mantle of its contribution to the Revolution. Hence, from Soviet beginnings, land-bound doctrine held the development of the fleet locked in a vise. Although the exploits of the Soviet Navy during World War II were anything but spectacular, that war nonetheless generated one watershed for Soviet sea power—not because of the role played by the Russian Navy, which was relatively insignificant, but for quite different reasons.

From that conflict, the USSR emerged as a superpower—one which rose, Phoenix-like, from its own ashes. Before the war's last shot had been fired, Stalin realized he would no longer be confined to trying to sell communist ideology around the world through the Comintern and its "front" organization adjuncts. He would now possess the military strength and geopolitical position from which he could impose his views on other peoples, even as the Bolsheviks had done within the Soviet Union. Losing little time in employing these new-found advantages, he had by 1948 swallowed up all of Eastern Europe, stood on the shores of the Persian Gulf, and contested with the Western powers for control of Austria.

But the reach of armies—even ones as powerful as those which Moscow then disposed—is limited by coastlines. Despite the existence of this fundamental maxim—one that true maritime nations have known and understood for centuries—the Kremlin, trapped by its Red Army mentality, still only dimly perceived it. Not until the accession to power of Nikita Khrushchev did the maritime tide in the USSR really begin to turn. Even then there was a considerable amount of irony in the shift, inasmuch as it was Khrushchev who sacked Admiral Kuznetsov, Stalin's head of the Soviet Navy. The new Soviet dictator replaced him with Admiral Sergei G. Gorshkov, who would eventually frustrate Khrushchev's naval dictums—dictums one can fairly assume were heavily influenced by Stalin and reinforced by Khrushchev's own wartime service with the Red Army.

In one of the new Russian Premier's first moves, he ordered scrapping of the Soviet Navy's on-going cruiser construction program. In his mind, the weapon of the future simply had to be the submarine, armed with missiles of various types. One must presume that Khrushchev hired Admiral Gorshkov to implement this kind of naval strategy and, equally important, to sell it to the rest of the Soviet naval leadership. But Khrushchev reckoned without the sea power acumen and political expertise his new man brought to the job. Today, more than twenty years later, it seems clear that Sergei Gorshkov set out at the very beginning of his tenure to correct the land-bound thinking of the entire Kremlin hierarchy—Nikita Khrushchev included. The main thrust of his argumentation maintains that

> The constantly growing sea power of our country ensures the ever wider use of the immense resources of the World Ocean in various branches of the national economy. . . . The use of resources of the World Ocean in association with the further development of science in this field is opening up new directions in the economic and political integration of the socialist countries, expanding the sphere of their international cooperation and raising the prestige of the Soviet state in the international arena. . . . The growing sea might of our country ensures the successful conduct of its foreign policy, helps constantly to widen trading, merchant, scientific and cultural links with other countries and to strengthen the constructive cooperation of states with different social systems and it places in the hands of our people a most important means for fulfillment of its historic

mission—the constant expansion of economic aid to all countries which have begun independent development.[1]

A minimum of reading between the lines renders this passage a rather clear statement of Soviet maritime objectives as conceived by the senior Soviet naval commander.

In his campaign to justify sea power within the walls of the Kremlin, Gorshkov has been singularly successful. Certainly a persuasive case can be made—based on his public writings—that he has long advocated the importance of conventional sea power, especially to any nation harboring pretensions of global leadership. Moreover, one can reasonably surmise that confrontations such as the 1956 and 1967 Middle East wars, Lebanon in 1958, and the 1962 Cuban missile crisis gave him a host of solid arguments which he knew would carry great weight in the Kremlin and which he used with telling effect.

There can be little doubt that during the days of late October 1962, for example, Nikita Khrushchev found himself backed into a military and political corner from which there appeared to be no exit except that of abject surrender or the nuclear holocaust the entire world hoped to avoid. That particular corner was fashioned by the U.S. Atlantic Fleet when it clamped a blockade—at the time euphemistically dubbed a "quarantine"—around Cuba. Diplomatic and international niceties aside, however, it was a blockade. And, significantly, the Soviet Navy had very little relevant power with which to offset the American action. The contrast between the blockade of Berlin—successfully vaulted by American and Allied air power—and the "quarantine" of Cuba, which a relatively impotent Soviet Navy could not penetrate, certainly could not have escaped the notice of the Kremlin.

The major lesson to be learned by the Soviets from previous crises, with all their moves and countermoves, and hammered home once again by this one, was that, distant from one's own shores, a basically land-oriented nation like the USSR can do precious little to control the outcome of a confrontation with a major maritime power. In such circumstances, even a newly-arrived superpower must be able to deploy significant afloat strength if its views are to be respected—and, above all, heeded. If Gorshkov did not repeatedly make this crucial point in the halls and council chambers of the Kremlin over the years, he was a fool. And it

---

[1] S. G. Gorshkov, *The Sea Power of the State* (Annapolis, Md.: Naval Institute Press, 1979), pp. 57-58.

strains credibility to think that a fool could cling to the leadership of the Soviet Navy for more than two decades.

Based on the evidence now at hand, one must conclude that a fundamental shift in Moscow's geopolitical thinking took place in the late 1950s. Until that time, the USSR had been exceedingly busy securing its borders, consolidating internal control of the new nation, surviving World War II, then reaching beyond its frontiers into areas readily accessible to the triumphant Red Army.

In the days and months immediately after the war, transforming the military conquest of Eastern Europe into lasting political gains and ensuring security against any new forms of attack from that direction were clearly primary Soviet objectives. Even in those early days, however, Joseph Stalin did not overlook the vistas which his new-found geographic position and fully mobilized military power opened up. It took some very tough U.S. and Allied talk, for example, as well as considerable flexing of the few military muscles remaining to the United States and the United Kingdom to get the Russian Bear out of Austria and Iran.

To be sure, American nuclear weapons—still perceived as a monopoly in those halcyon days—proved to be the trump card. Absent this factor, it is unlikely that Stalin would have been dissuaded by the rapidly dwindling American and British conventional military strength from absorbing all of Europe. As a matter of fact, it must have been a very gleeful Joseph Stalin who watched Winston Churchill's prophetic May 1945 question become reality: "What will be the position in a year or two, when the British and American armies have melted . . . and when Russia may choose to keep two or three hundred (divisions) on active service?"[2]

In subsequent years, only President Harry S. Truman's economic and military aid program—led by his military hero and Secretary of State, George C. Marshall—frustrated Soviet designs in Greece. And the postwar history of Berlin is all too well known. But if anyone thinks that the Kremlin has always been infallible in conducting its foreign policy, the Berlin blockade should certainly dispel that myth.

As an embryonic European Union struggled somewhat unsuccessfully to pull itself together and fashion a meaningful defense against the manifest Soviet threat which had just replaced that of a predatory Nazi Germany—the latter challenge being one the European nations had all just exhausted

---

[2]Winston S. Churchill, *Triumph and Tragedy* (Boston, Mass.: Houghton Mifflin, 1953), p. 573.

themselves defeating—the Russians set up the Berlin Blockade. For its efforts, Moscow got not Berlin and a united communist Germany, but the North Atlantic Alliance—and its military counterpart, the North Atlantic Treaty Organization.

By imposing the blockade, the Soviets drove the European Union to vault the Atlantic Ocean where it enlisted in its ranks the one nation with which the Kremlin could not at that time compete: the United States. Hasty formation of the Warsaw Pact, followed by almost thirty years of Soviet scheming and laboring to dismantle NATO offers abundant testimony to the often unrecognized but periodically demonstrated ability of Moscow to shoot itself in the international political foot.

Berlin in 1948; Hungary in 1956, when only the ill-conceived British-French-Israeli adventure at Suez drew world attention away from the Soviet tanks clanking through the streets of Budapest; the invasion of Czechoslovakia in 1968, which shocked NATO's member states out of their euphoria over the newly-advertised detente they were falling all over themselves to embrace—all testify to the less than astute manner in which Soviet foreign policy is periodically executed. Despite these major international reverses—yet another occurred in 1972 when a gambling Anwar Sadat ejected the Soviets from Egypt—Moscow has nonetheless kept trying. And it was this early campaign to project Soviet influence beyond the boundaries of the so-called heartland that laid the groundwork for the present, far more successful drive for maritime predominance.

Initially, Stalin used arms grants as his opening political wedge. Indonesia, flanking the Strait of Malacca, comes to mind as an early example of a country and an adjacent maritime choke point targeted by the Kremlin—although at that time the Soviet motivation had not yet focused on maritime hegemony. The Russian dictator came very close to achieving success in Jakarta. Two disparate factors, however, intervened to decisively frustrate his aims in this huge archipelago. The influence of Communist China, stemming for the most part from the large indigenous Chinese population in the country, gradually shifted the communist movement to Peking's side of the fence—another direct consequence of the Sino-Soviet split. But, when the showdown came in Indonesia, Western interests were ultimately salvaged by an entirely different factor which most of the American people and the bulk of the U.S. Congress have never recognized.

Most Indonesian military officers had been trained in the United States. After spending a few months to a year or more in this country, they returned to their homeland disabused of the notion that Americans were arch-imperialists, as Russian and Chinese communists constantly tried to make them out to be. Some Indonesians, on extended training or educational tours, were accompanied by their families, which also returned home with generally favorable impressions of Americans and their social and political institutions. Thus, when President Sukarno and the Indonesian Communist Party (PKI) tried to take Indonesia into the communist camp—albeit Chinese oriented by this time—virtually all of the military leaders of the nation came down strongly in support of national independence and friendship with the West. The monies the United States had invested in their training and education had paid handsome international political dividends, which few American citizens or their Congressmen had foreseen. Once more, the Soviets failed to attain their objective.

While the Kremlin hierarchy must have been dismayed by each of the setbacks described above, one would be foolish to assume that the Soviet leadership learned no lessons from them. One of those lessons—no doubt clearly perceived and astutely advanced by Gorshkov—must have been that, beyond the heartland, the reach of the Red Army always ran afoul of salt water. Gorshkov's remedy was a simple one: Build a blue-water fleet capable of projecting power and influence around the world to regions far beyond Soviet borders. From the late 1950s onward, Moscow appears to have been pursuing this precise course of action.

Furthermore, it seems evident that in planning their campaign the Soviets tore a page directly out of the book of Imperial England. A close study of the map today reveals that the USSR is busily making international mischief in the immediate vicinity of many of those major maritime funnels which Great Britain controlled until recent years. Moscow is not exploiting them all, to be sure, but certainly enough of them to suggest strongly that they have taken aboard a lesson of the first magnitude: When the international chips are down—in an environment of true strategic deterrence—he who commands the world's shipping choke points, controls this interdependent world in which we now live (Sir Halford J. Mackinder notwithstanding).

Having apparently subscribed to this basic precept, the Soviets immediately embarked on the previously described naval expansion program and, under Admiral Gorshkov's tutelage, the result to date has been the emergence of a deep-water fleet of astonishing size and impressive capability. Today, modern Soviet warships operate routinely in every one of the globe's oceans, and in every instance they have clearly come to stay.

The extent of the Soviet penetration of distant oceans is starkly revealed by Graph 1. The vertical standpipes portray Soviet Navy ship-days spent out of home waters for selected years from 1965 to the present. Annual totals are further broken down by region. The dramatic rise which occurred in the late 1960s directly reflects the rapid increase in the numbers and sophistication of warships in the Soviet fleet which occurred after Gorshkov became head of the Navy. The graph also indicates the rapidity with which Moscow began to exploit this expanding outreach of the Soviet Navy. The foregoing general conclusions are particularly apt with respect to the Indian Ocean where, commencing in 1968, the USSR sent its Navy to establish a permanent presence in that distant, but critical body of

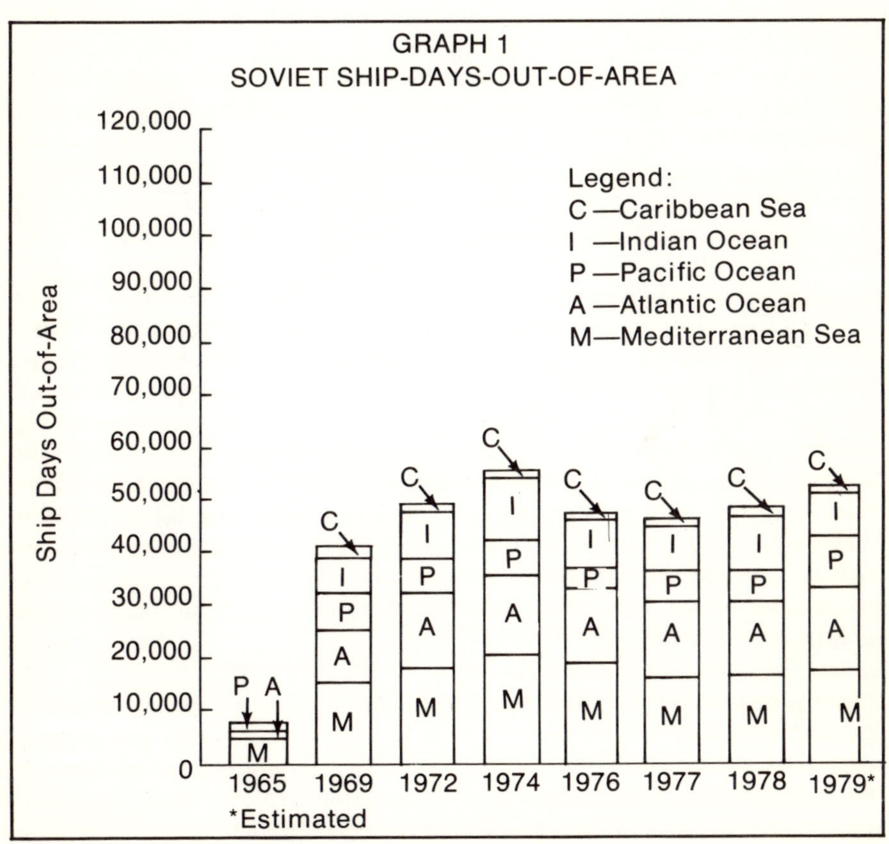

water. Indeed, men-o'-war flying the Hammer and Sickle have been there in strength ever since.

No one could have been more aware of that presence than I was during the early days of October 1973 when the fourth Middle East war erupted and I found myself with three ships in the entire ocean while the Soviets disposed some two dozen. Not until a task force, built around the old and comparatively decrepit aircraft carrier *Hancock*—in the words of then U.S. Ambassador to India Daniel Patrick Moynihan—"huffed and puffed its way through the Strait of Malacca," did I begin to feel a little more comfortable.

Far to the north in the Mediterranean, my friend and cohort, Vice Admiral Dan Murphy, must have felt much the same way as I did, as he watched the Soviet squadron in that strategic sea grow swiftly to a strength of more than 90 ships, while his own famed U.S. Sixth Fleet could muster but 70. This was a far cry from the Lebanon crisis of 1958; and the numbers of American and Soviet ships in 1973, widely reported in the various news media, did not escape the notice of littoral nations in the Mediterranean or in the distant Indian Ocean. That Moscow now understood the value of sea power could no longer be seriously questioned; nor could there be much doubt that the Kremlin, having achieved a considerable degree of maritime maturity and having acquired the sea-going hardware to go with it, was ready to put that knowledge and capability to use.

But while these events were unfolding, the Soviets continued to pursue a far more insidious campaign, one little noticed in the West in general and in the United States in particular. As a matter of fact, despite the USSR's worldwide incursions of recent years, this campaign is not widely perceived in America even to this day.

Moscow—now boasting the maritime power inherent in an ocean-going, blue-water navy, along with huge merchant, fishing and oceanographic fleets—is well on the way to acquiring that which Great Britain possessed almost a century ago: domination of every major shipping focal point on the face of the globe. London had to relinquish control of these critical locations as a result of two consequences spawned by World War II: fiscal exhaustion and the demise of colonialism. An infinitely greater tragedy, however, appears to be in the making. For, that which the British found themselves forced to give up, the United States seems determined to permit the Soviet Union to acquire without challenge. This nation is, in fact, voluntarily leaving the field to communist exploitation—a decision which, unless reversed, will assuredly redound to its great detriment.

5. The Pattern of Soviet Advances

To be sure, one certainly cannot argue that Moscow has been universally successful in its reach for the sea. While the various men in the Kremlin have made significant gains across the years, they have also suffered some stunning defeats. Nevertheless, the main thrust of their policies remains. And their current maritime offensive must be viewed as one which, based on the evidence to date, is ultimately likely to succeed unless it is countered by the United States. It is this prospect which bodes considerable ill for the West in general and the United States in particular. The Soviet drive raises two crucial questions: First, where and to what degree has the USSR succeeded to date? Second, what are Moscow's future targets?

The Caribbean Sea and Atlantic Ocean

Starting on America's own doorstep, one finds the Soviet Navy firmly entrenched in the Caribbean Sea. Operating from Havana and Cienfuegos, units of the Red Fleet—sea and air—are in excellent position to sever the shipping lanes which crisscross that important body of water. Despite the huge financial drain that Fidel Castro's regime has imposed on the USSR, the strategic value of the island's naval operating facilities is so great that the Kremlin must be convinced that it has a very good bargain indeed. Moreover, since the use of these facilities in exchange for heavy Soviet subsidies is now accompanied by the recent employment of Cuban surrogates in other parts of the globe—most especially in Africa—the bargain must seem better than ever.

Unlike the British, who acquired numerous island and continental colonies which gave them redundant control of that sea, Moscow has so far settled for only one central location: Cuba. From the time of Castro's seizure of power in 1959 until the ill-fated Bay of Pigs invasion in 1961, few Americans gave much thought to the island. But subsequent to the 1962 Cuban missile confrontation between the United States and the Soviet Union, Moscow's machinations on that island have been seen in the United States primarily in light of the strategic nuclear threat. Comings and goings of Soviet surface warships and naval aircraft have been little remarked outside American naval circles. Only when the Soviets have made periodic moves designed, one must presume, to test the alertness and determination of the current Administration in Washington—by sending ballistic-missile-equipped submarines and their tenders on visits to Cuban ports, or by making ostensible efforts to establish permanent operating bases for those units—have their actions caught the attention of the American press and, thereby, of the American people.

In 1970, for example, many of the 1962 missile-crisis tensions returned center stage when the Soviets sailed a submarine tender into Cienfuegos on the island's southern coast, and American U-2 reconnaissance flights revealed construction activity ashore of a type clearly connected with submarine operations. The tender departed only after the U.S. Government let it be known publicly and forcefully that it viewed the tender's presence, along with the associated work ashore, as a direct and grave violation of the agreement by which the two countries had resolved the 1962 confrontation. Construction ashore, however, apparently continued.

Again, in 1972, Moscow seemed bent on sampling the political atmosphere when the Kremlin ordered the first ballistic-missile-firing submarine, a diesel-powered Golf II, into Cienfuegos. Washington's hackles rose once more. If the ultimate Soviet aim behind these moves was eventually to establish a forward operating base for the USSR's Yankee-class, nuclear-powered, strategic submarines—similar to the American practice at Holy Loch, Scotland, and Rota, Spain—Moscow must have concluded that Washington would not tolerate it, because no such visits to Cienfuegos have taken place.

While these Soviet naval forays captured nationwide attention in the United States, comparatively little domestic notice was taken of other Soviet fleet deployments to Cuba. Appearances of naval aircraft and the most modern ships in the Soviet Navy had, by this time, become routine occurrences in Cuban waters, the Caribbean Sea, and the Gulf of Mexico. Yet, no one in the United States seemed to be much concerned about a fundamental fact of maritime life emphasized by this continuing Russian presence: From Cuban ports and airfields, Soviet naval and air forces could interdict all of the main shipping routes connecting South America and Panama with the Gulf and East Coast ports of the United States, as well as those sea lanes extending from the Panama Canal and the Gulf of Mexico to the harbors of Western Europe.

Taken alone, the Soviet presence in the Caribbean is of sufficient significance to warrant serious apprehension in the United States, but when coupled with other Soviet initiatives undertaken in the Atlantic Ocean basin—especially in the South Atlantic—the larger pattern thus disclosed assumes even more ominous dimensions. Nevertheless, because the threat thus posed by past Soviet actions was neither as spectacular nor as immediate as that presented by the sea-launched ballistic missile, it was largely ignored by most Americans.

By any measure the United States might choose, the Soviet politico-maritime campaign in the Caribbean must be judged eminently successful to date. Moreover, the importance of the Cuban connection is clearly multiplied when one looks across the Atlantic to the northeast.

Soviet warships and maritime patrol aircraft now routinely stage out of Northern and Baltic Fleet bases, transiting the broad reaches of the Atlantic Ocean to supplementary bases in Cuba. Enroute to and from the Soviet Union, their courses and flight paths take them across all of the shipping routes lacing the North Atlantic, as well as along the entire eastern seaboard of the United States. The use of airbases in Cuba has afforded the Soviet Navy greatly extended coverage of the North and Central Atlantic, together with far greater time on station in those mid-ocean regions being patrolled. As for Soviet warships, Cuban harbors provide them with all the advantages of forward deployment—advantages long known in Western fleets.

Then, if one turns southeastward from Cuba—as the Soviets, flushed with success following the Cuban experience, obviously did—one discovers the Soviet Navy fairly well established along the West African coast. Over the past decade and a half, Moscow has exerted considerable effort to infiltrate the littoral nations in this part of the world, particularly the Congo, Guinea and, most recently, Angola. From seaports such as Conakry, Pointe Noire, Luanda and Mocamedes, those same Russian warships and long-range maritime patrol aircraft today maintain a presence in the South Atlantic, and the Kremlin's efforts here must be judged nearly as successful as those in the Caribbean. And, as in the Caribbean, these South Atlantic operating facilities place Soviet naval prowess astride major world shipping lanes leading to North America and to Europe. In this connection, it is important to recognize that—some flow of critical minerals and metals from West Africa aside—the main significance of the South Atlantic is the fact that it serves to connect the North Atlantic to the vital Indian Ocean.

If one now draws on a map a series of lines from Luanda and Conakry to Cuba, thence to the North Cape, a chilling picture materializes: The USSR today has the naval forces and operating bases necessary to interdict every significant shipping lane in the North and South Atlantic Oceans, as well as all of those in the Caribbean Sea. This, then, is what the Soviet Union has achieved in the Atlantic; and its posture raises yet another crucial question: What about the future of this part of the world, particularly the future of Southern Africa?

The Soviet-Cuban intervention in Angola—and the failure of the United States and other Western nations to react thereto—assumes crucial importance when one examines closely what is currently happening throughout the rest of Southern Africa. For, on the opposite side of that troubled continent, Russian aid—swiftly and massively mounted—transformed yet another of Portugal's colonial holdings, Mozambique, into a Marxist stronghold. Here, long before the communist hand revealed itself in Angola, this East African nation had fallen to Soviet-backed rebels. Thus, Moscow now not only enjoys the use of maritime operating facilities in these two countries, but has been supplying radical guerrilla forces which conducted a long campaign of conflict and terrorism against the white minority government in Rhodesia and finally succeeded in bringing about its downfall. The marathon British effort to negotiate a solution to the Rhodesian problem has now brought to power Robert Mugabe's Marxist guerrilla faction. Thus, today a glance at the map of Southern Africa shows that a belt of Marxist states stretches across the continent from the Atlantic to the Indian Ocean. The portents for such states as Namibia, Zambia and Botswana aside, it should be clear that the pressure on South Africa will now become severe.

What conclusion can one draw other than that the world is witnessing a concerted Kremlin drive for dominion over the entire southern portion of the African continent? If successful, this campaign would place Soviet naval power athwart the extremely important shipping route at the Cape of Good Hope—the "windy corner" of the Middle East oil trade. It seems obvious that here the Kremlin is reaching for control over yet another of the pink-hued choke points on those high school maps, this one bounded on the north by the African coast and on the south by the gale-force winds and mountainous seas of the "Roaring Forties" and "Screeching Fifties." It hardly need be said that a Soviet triumph such as this, comparable to earlier successes in the Caribbean Sea and West Africa, would provide Moscow with something that a decade ago would have been inconceivable: a double handhold around the industrial throats of Western Europe and the United States.

The Horn of Africa

Doubling the Cape of Good Hope and entering the Indian Ocean, a survey would find the Soviets busily boring away throughout the region. In East Africa, a little more than 600 sea miles north of Mozambique, Somalia offers a classic example of Soviet penetration of a target nation using subversion and military aid.

With the overthrow of the ruling government in 1969, and less than a year after the establishment of a permanent Soviet naval presence in the Indian Ocean, Russian-backed Major General Mohammed Siad Barre seized power and almost immediately opened the nation's ports to the Red Fleet. Astutely selecting the relatively undeveloped port of Berbera on the Horn of Africa, Moscow set to work building a naval operating base commanding the Bab el Mandeb, the southern entrance to the Red Sea. (Berbera, incidentally, lay in British Somaliland until that territory was conquered by Benito Mussolini's legions early in World War II.) Concurrent with the move into Berbera, the Soviets were heavily engaged in aiding the newly established People's Democratic Republic of Yemen with its superb port of Aden, still another colonial possession of Britain, which had held it since 1839. Just as England had done in the previous century, the USSR now began to fashion a maritime clamp on the southern entrance to the vital Red Sea.

According Berbera high priority, the Soviets soon had the harbor dredged, new piers constructed, and a barracks ship as well as a sizable drydock installed. Ashore, a naval communications station and antenna field were built, together with an airfield and, eventually, a closely guarded missile handling and storage facility. Ships of the United States Middle East Force regularly observed Soviet warships going to and coming from Berbera.

The U.S. Government followed the development of this new naval base with growing concern and eventually began to expose its existence to the rest of the world. The Government's effort was met by heated denials from Moscow and Mogadiscio along with expressions of disbelief by some members of the U.S. Congress. To prove the validity of its assertions, the Administration finally released to the public a series of reconnaissance satellite photographs revealing the activity in minute detail. Still denying the existence of a naval base, Barre invited a group of Congressmen to come and see for themselves. Despite efforts to conceal the true nature of Berbera's military installations and a flat refusal to permit the visitors inside the missile facility or the communications station, the Congressional delegation came away with the firm conviction that the Administration was indeed telling the truth.

Even as this comic-opera exchange was taking place, Moscow was casting a covetous eye on another African prize. To the north, a radical Marxist military junta had seized the reigns of power in Ethiopia from the fallen Emperor Haile Selassie, immediately moving to eject the Americans from the country and entice Soviet arms and assistance. The rapidity with which Moscow responded provoked inevitable speculation that the Kremlin had once again shot itself in the political foot.

Ethiopia and Somalia had long been locked in a bitter dispute over ownership of the Ogaden, a vast stretch of territory in southern Ethiopia. Peopled by several tribes which historically roamed the area without regard for such things as artificial international boundaries they could neither see nor comprehend, the Ogaden had also recently been the site of extensive exploration for petroleum. None had been discovered, but some gas deposits had. This only heightened the competition between the two countries for this stretch of Ethiopia wherein the bulk of the population comprised Somali tribesmen. Like many of the flashpoints in this part of the world, the Ogaden constituted a bubbling stew of contradictions and divided loyalties, leading to international confrontation. It was into this morass that the Soviets stepped when they agreed to re-equip the Ethiopian Army.

Many observers around the world grinned and sat back, waiting for the Soviet leaders to fall off their hastily erected tightrope. It seemed clear that the Kremlin had embarked on an impossible mission—trying to be friendly to each of these mutually hostile nations. Indeed, it is difficult to explain the haste with which Moscow, already established in Somalia, leaped to the aid of the new Ethiopian regime—until one examines all facets of the equation.

In recent years there have been persistent rumors of Somali irritation with the Soviets and their presence in that country, of Saudi Arabian pressure on Mogadiscio to expel them, and of conduct on the part of Russian personnel in Somalia which served only to compound the other problems. Thus, it seems quite probable that in spite of the massive investment Moscow had made in this primitive East African nation, it came to believe that Soviet tenancy would nonetheless remain rather tenuous. Perhaps the leaders in the Kremlin considered the Soviet-Somali Friendship Treaty to be a document with a limited life expectancy, and that whatever they did in neighboring Ethiopia would make very little difference.

In some circles it is fashionable to credit Moscow with possession of a carefully-devised, long-term strategic plan for global domination. Those holding this view contend that each and every Russian move around the globe demonstrates not only the existence of such a plan, but its systematic prosecution as well. Another school of thought disputes this thesis, maintaining that, while Moscow may have a general notion of where it is taking the Soviet Union, its initiatives are more ad hoc in nature. On balance, it would appear that these events in and around the Horn of Africa suggest that the latter view is the more plausible.

They demonstrate, rather well, the notion that the USSR is more of an opportunist than an innovator in the implementation of foreign policy, a predatory nation quick to strike at a target of opportunity rather than one expert in creating such opportunities. No one could realistically contend that the Russo-Ethiopian connection evolved from careful, long-range, strategic planning within the Kremlin. It most certainly did not. Obviously, a sudden break in the historic political patterns in the region presented Moscow with an unforeseen opening, which it seized with alacrity. This is not to argue, however, that the Soviet Union has no overall plan for East Central Africa. If one recalls previous Soviet actions in Egypt, the Congo, Guinea, Sudan, Somalia, Angola, Mozambique and elsewhere, the notion that the Soviets harbor no extensive ambitions in the African Continent must be regarded as ludicrous.

Certainly, when the U.S. Communications Station at Asmara in Eritrea was shut down, and the American military mission expelled by the new Marxist rulers in Addis Ababa, Moscow must have at once perceived a golden opportunity to exploit yet another African country. After all, Mohammed Siad Barre had not, at this time, ordered any overt military move against the long-coveted Ogaden region of southern Ethiopia.

Leonid Brezhnev and his advisers may, therefore, have assumed that in those circumstances they could successfully walk that international tightrope, thus enjoying the best of both worlds. If so, the sudden Somali military assault—guerrillas at first, openly backed in very short order by regular Somali armed forces—must have come as a most unwelcome development. Once the battle was joined, of course, the Russians found themselves caught in a next-to-impossible position, backing both sides of a bitter local conflict—an untenable posture in the long term. Whether the men who rule the Kremlin believed that such an international balancing act could be pulled off is not terribly relevant. One must conclude that other, more important considerations played a decisive role in the Soviet decision to swiftly espouse the new Ethiopian regime.

In Ethiopia, an avowedly Marxist junta had seized power from a tottering and increasingly discredited emperor, embraced communist tenets with well-publicized enthusiasm, and then set about converting the entire country to those beliefs. Moreover, this ancient kingdom is ideally situated to provide Moscow with a base for subversive incursions throughout the central portion of the continent.

Abutting Sudan, Kenya, Djibouti and Somalia, as well as being but a stone's throw from Uganda—a potential highroad to Tanzania and Zaire, the latter already beset by a communist-inspired insurgency mounted from Angola—Ethiopia surely appeared in Moscow as a perfect staging area for thrusts into all of Central Africa. This prospect alone would certainly make the gamble worthwhile.

Furthermore, if the Soviet Union was concerned that continuing Saudi Arabian pressure on the Moslem leadership of Somalia to cut its ties to Moscow would eventually bear fruit, Ethiopia would offer a most attractive alternative. Finally, complete loss of the base at Berbera and the airfield at Mogadiscio would not leave the Soviet Navy without shore-based support in the region, given the close political and military bonds between the USSR and South Yemen. After all, continued access to the excellent harbor and shore facilities left behind at Aden by the British, along with use of the anchorages off the South Yemeni-owned island of Socotra, would still permit Soviet warships and maritime patrol aircraft to dominate the southern approaches to the Red Sea, even should the USSR be ejected from Somalia and the Ethiopian adventure fail completely.

On the other hand, if the latter effort were to succeed—including an Addis Ababa victory over the Eritrean rebels, thereby assuring the Soviet Union access to the naval base and port facilities at Massawa—Moscow's naval influence would be extended all the way to the southern terminus of the Suez Canal. From the Kremlin's point of view, this prospect was likely irresistible. The location of Massawa, situated as it is in Eritrea, goes far toward explaining the overt Cuban assistance to the Ethiopian junta in its campaign to put down the long-lived rebellion in that strife-torn province.

Thus, in late November of 1977, when Mohammed Siad Barre tore up the friendship treaty with the USSR, the Soviets dutifully departed, taking their massive drydock from Berbera to Aden. One suspects that the communications equipment and the missiles previously stored at Berbera also made their way to Aden. Despite its reversal in Somalia, the Russian Navy continued to ply the waters adjacent to the Horn of Africa, and the aforementioned benefits of an ultimate Marxist triumph in Ethiopia remained eminently achievable. Moreover, extreme reluctance on the part of the United States to ship military aid to the Somalis during and after the Ogaden war—ostensibly because of long-held Somali claims to portions of Kenyan territory—seemed to guarantee that Mogadiscio would not move promptly into the Western camp.

The Persian Gulf and Arabian Peninsula

To the east, the sea-going Hammer and Sickle has moved regularly in and out of Iraqi ports, thanks to another of those ubiquitous "treaties of friendship and cooperation," which provides Moscow a maritime anchor at the head of the all-important Persian Gulf and access to naval support facilities. Close ties with Baghdad began with the overthrow of one government and the succession to power of another, the latter holding strong Marxist-Leninist leanings.

The downfall of the Hashemite monarchy in 1958—accompanied by the assassination of King Faysal II and Prime Minister Nuri as-Said—was followed by a succession of coups and countercoups, ultimately bringing Hasan al-Bakr and his Revolutionary Command Council to power in 1968, with the Iraqi Communist Party participating in a coalition government. Moscow at once went to work. Trading on the results of the 1967 Arab-Israeli war, the Kremlin moved to attract Iraq into the Soviet orbit and, by 1972, had negotiated a "Treaty of Friendship and Cooperation" with the Baghdad Government. Persistent rumors are extant that the agreement also encompassed a secret military protocol. In any event, the Red Fleet was not far behind.

By 1973, Soviet warships were calling regularly at Basrah, and Moscow had undertaken to construct a wholly new Iraqi naval base at Umm Qasr on the shores of the Persian Gulf. As the months passed, Soviet use of the facilities in Iraq increased both in frequency and length of stay. Repair ships, for example, spent extended periods in Basrah providing upkeep and maintenance for Soviet minesweepers and intelligence collectors, after which the latter would sail down the Gulf to the Strait of Hormuz where they would linger, monitoring the traffic—naval and commercial—transiting the waters of that narrow choke point. Moscow was not unaware of the strategic significance of the Strait, the passage through which moved all outward-bound Persian Gulf petroleum—from Iran, Kuwait, Saudi Arabia, Bahrain, Qatar, and the United Arab Emirates—and conditions ashore seemed fortuitously ripe for exploitation.

Across the southeastern end of the Arabian Peninsula lies the Sultanate of Oman, beset in those days by a little known rebellion mounted in the western province of Dhofar, and sustained by Soviet aid channeled through the neighboring Marxist-ruled country of South Yemen. Funneling arms and supplies through this Arab surrogate, Moscow sought to ensure the success of the Dhofar rebels.

This essentially unpublicized insurrection in a remote part of the globe was poorly understood, not only in the United States but throughout most of the rest of the world. Nonetheless, because of an obscure geographic fact, the Dhofar rebellion incorporated grave dangers for the Western world. This geographic fact also explains the intense Soviet interest.

Jutting like a spur into the Strait of Hormuz—the sole exit from the Persian Gulf—is the Musandam Peninsula. Entering and exiting the Gulf, main shipping lanes run between this projection and the nearby Quoin Islands. Sea-borne traffic could thus be subject to complete domination, or interdiction, by modern military weapons emplaced ashore on this spit of land. Outbound ships—most significantly, heavily laden tankers—hug the shores of the Peninsula itself.

The fact of geography? Although separated from the bulk of the Sultanate by mini-states of the United Arab Emirates, the tip of the Musandam Peninsula is nevertheless Omani territory. Thus, he who controls Oman also controls this desolate bit of land, the traffic moving into and out of the Persian Gulf—and, ultimately, the supply of the region's incredible petroleum reserves to the rest of the world. In this light, Moscow's avid interest in bringing a Marxist-Leninist regime to power in Oman becomes manifest, as do the actions of the Soviet surrogates—Iraq and South Yemen—in support of the entire "Arabian Gulf Liberation" movement.

The implications for oil supplies to the Western world, if the Kremlin were to achieve that sort of dominion over this choke point, should be obvious to the most casual observer. It most certainly was to Great Britain and Iran. Both countries clearly perceived the danger posed by a radical government in control of Oman and moved to help Sultan Qaboos put down the insurrection—first Great Britain and then Iran. Because of their century of control over the Persian Gulf nations and the Sultanate of Oman, the British were already on the ground and involved. The Sultan's armed forces were British-led, London had troops stationed in the country, and the Royal Air Force operated a sizable military airfield on the island of Masirah off the southern Omani coast.

Despite this assistance, the Sultan made little progress against the insurgents for several years. Then the Shah of Iran, who also had an immense stake in free passage of Iranian oil through the Strait of Hormuz, committed his own army, air force and naval units to the campaign. In 1975, thanks in part to this additional assistance, the rebellion was finally crushed and most of the leaders captured. It should be noted, however, that the rebellion currently is quiescent rather than dead, and recent events in Iran make it likely that it will be revived.

One is constrained to observe at this point that domination of shipping choke points is not the principal driving force behind Soviet state policy in this particular segment of the globe. If past and present Soviet efforts in Egypt—ultimately a failure, of course—Sudan, Ethiopia, Somalia, Oman and Iraq are considered, it seems evident that the prime Soviet objective here is to draw a noose around the Arabian Peninsula with its gargantuan oil reserves. Adding to this mosaic Moscow's post-World War II attempts first to occupy Iran permanently and later to engineer the rise to power of a friendly regime in Teheran, it becomes clear that control of the world's major energy supplies must be the primary Soviet aim. In the case of the Persian Gulf, the choke point campaign would appear to be a matter of back-up insurance.

The Indian Ocean

On a southerly course from the Persian Gulf, enroute to the Pacific Far East, one encounters a struggling ex-British colony off the tip of the Indian subcontinent: Sri Lanka. Impoverished, plagued by problems of all sorts, neither this tiny nation's diminutive size nor its position in the international power pecking order would appear to act as a magnet, attracting the attention, let alone the blandishments, of a major world power. Yet, Colombo, its capital, is home to an incredibly large military mission in the Soviet Embassy. Moreover, Moscow recently bent every effort to assume the role of benefactor when the embryonic Sri Lankan Navy went looking for a small ship to serve as the nation's fleet flagship. In 1976, that ship arrived, a *Petya*-class anti-submarine frigate donated—no other word adequately describes the transaction—by the USSR.

Despite the seeming unimportance of Sri Lanka, one need not look very hard to find plausible explanations for the Soviet largess: First, of course, is the island's geographic location. The waters south of Sri Lanka constitute another "windy corner" among the globe's shipping lanes; they constitute a veritable crossroads in the Indian Ocean. The main shipping lanes exiting the Red Sea and the Persian Gulf pass just south of Sri Lanka enroute to Pacific Far Eastern destinations as do those running from the Cape of Good Hope to the western entrance to the Strait of Malacca. It thus becomes exceedingly clear why the Soviets have cast an eye on the tiny nation. Great Britain fully understood the importance of this strategic position and built an incomparable naval base at Trincomalee on the island's eastern coast. One can be reasonably certain that this facility also plays a significant part in Moscow's abiding interest in Sri Lanka. When these realities are coupled with the evidence of

Russian interest in strategic maritime foci elsewhere in the world, Sri Lanka's attraction becomes less inexplicable.

Roughly 500 nautical miles southwest of Sri Lanka sit the Maldives, a string of tropical atolls stretching north to south across eight degrees of latitude. Comprising some 2,000 islands, grouped into about 20 atolls, the Maldives—one of the least developed countries in the world—were granted their independence by the British in 1965. Again, here is an impoverished mini-state, struggling desperately to survive since being cut loose from England's colonial apron strings.

While lacking practically all of the attributes necessary to become a viable independent nation in the modern world, the Maldives do boast one asset capable of attracting international attention. Bequeathed to them by their late British overlords was an excellent military airfield on the island of Gan in the Addu atoll at the southern end of the archipelago. An aerial refueling and staging base, Gan is yet more. Maritime patrol aircraft with a range of two thousand miles—a modest capability in today's aircraft world—can cover the entire Indian Ocean from the Strait of Malacca in the east to the African coast in the west. Given the vast reaches of the Indian Ocean, the Maldives' geographic location thus assumes considerable strategic importance. Moreover, the Maldives stand like a sentinel, athwart the shipping lanes running through the Laccadive Sea, between the Persian Gulf and Red Sea, on the one hand, and the Pacific Ocean on the other.

Little wonder, then, that when the United Kingdom elected to pull back from "east of Suez," the Kremlin made a quick attempt to lease Gan's facilities. It was with a great sense of relief that Western nations—especially the United Kingdom and the United States—learned that the Maldives Government had turned down the bid. In light of the economic difficulties confronting the young country, however, there is no certainty that this or some succeeding group of leaders will continue to fend off the USSR. Here, for example, is a spot where Western economic initiatives—as opposed to military countermeasures, after the fact—can be used to frustrate Soviet maritime aims.

On the far side of the Indian Ocean lies the Indonesian archipelago where Moscow failed to secure a beachhead in the 1960s. Washing its northern shores is the Strait of Malacca which carries almost every ton of petroleum bound from the Middle East to Japan and nearly every other area of the Pacific, along with millions of tons of dry cargo. At the same time Moscow was seeking to pull Sukarno's Indonesia into the communist orbit, it was also fueling an insurgency in Malaya, one aimed at overthrowing British rule in favor of a Marxist government. Like the current Russian effort in the Horn of Africa, Southeast Asia can be viewed as another case of Soviet caution: hedge a bet by playing several possibilities at once. Fortunately for the United States, the Kremlin lost both of these wagers. One would be less than realistic, however, to assume that the Soviets have given up in this critical choke-point region.

Critics of the domino theory notwithstanding, developments in Southeast Asia since the American retreat from Vietnam have not been exactly encouraging. Two radical communist regimes now dominate this part of the Asiatic mainland: Vietnam and Cambodia. Continuing assaults against Laos—which many consider already lost—and Thailand augur ill for the cause of freedom in these tortured lands. The Sino-Soviet contest for hegemonic control of this part of the world has been most recently highlighted by the Vietnamese invasion of Cambodia, which toppled the Chinese-backed Pol Pot regime, and the Chinese assault on northern Vietnam in retaliation. The results of the latter armed confrontation were essentially inconclusive, however, and the ultimate outcome of events in Indochina still remains hidden in the mists of the future. Regardless of the eventual resolution of the Sino-Soviet conflict, it seems reasonably clear that this part of the world will eventually become completely Marxist in orientation. In either case—Soviet or Chinese success—it should be remembered that from this Indochinese power base, it is but a short bicycle ride—as the Japanese Army proved in the early days of World War II—to Malaysia and Singapore.

Should Moscow prevail, the outcome is not hard to envision: control of the Strait of Malacca by another Soviet-supported government. And, based on the patterns which have evolved in other areas of the globe—as well as those currently obtaining in Vietnam—one must assume that Soviet men-o'-war would quickly become a familiar sight in Singapore's incomparable harbor. Again, it should be noted that Singapore is yet another location on our familiar high school maps featuring those omnipresent pink or red colors, along with a huge naval base built by the British.

The pattern of Soviet initiatives in the Indian Ocean reveals rather clearly the determination with which Moscow has been utilizing its new-found maritime outreach. Only two months after Great Britain announced its decision to begin withdrawing its military and naval forces from "east of Suez"—a move which was to be completed by 1971—the first sizable Soviet naval squadron entered the Indian Ocean. Led by a *Sverdlov-*

class cruiser, the Russian force numbered five ships. Over the succeeding weeks, the ships called at some nine ports in seven countries. The facts attending this incursion are especially illuminating.

During the four months that the squadron remained in the Indian Ocean, its ships called at numerous ports including Madras and Bombay in India; Karachi, Pakistan; Colombo, Ceylon; Basrah and Umm Qasr in Iraq; Bandar Abbas, Iran; Mogadiscio, Somalia; and Aden. Moreover, the importance the Soviets attached to the cruise was signaled by the unprecedented trip of Admiral Gorshkov to India to meet the ships during their visits to Madras and Bombay. The excursion marked two significant firsts: establishment of a permanent Soviet naval presence in the Indian Ocean, and the beginning of a decade of close cooperation between the USSR and India, an association which—the invasion of Afghanistan notwithstanding—is likely to become close once again as a result of Indira Gandhi's return to power.

Little noted at the time, however, was the pattern of port calls, other than those to India. In light of subsequent developments, one must conclude that the visits to Colombo, Iraq, Aden and Mogadiscio were calculated investments in the future—gambles that at some time those port calls would pay dividends. Given the foregoing pattern and the speed of Moscow's reaction to the United Kingdom announcement, only the most naive observer would characterize the cruise and its timing as matters of coincidence.

The Pacific Ocean

Looking further eastward into the vast reaches of the Pacific Ocean, one finds the Soviet maritime design less well developed and, therefore, not quite so obvious. Perhaps this is because of the immense distances involved, both across its expanse and between the myriad islands which stud its surface. Nonetheless, here too the Soviets are active.

Aside from Korea and Indochina, there seems to be limited opportunity for Soviet machinations along the Ocean's western rimland on the scale evident elsewhere around the globe. The main barriers to Russian gains in the Pacific Far East center on the People's Republic of China and the U.S. Seventh Fleet. Both China and Japan have let it be known that they view the ships of this fleet as crucial to maintenance of the balance of power in the Far East. Following the trauma generated by President Carter's announcement, shortly after taking office, that he intended to withdraw all American ground forces from Korea, Secretary of Defense

Harold Brown hastened to assure the shaken Japanese that the United States would maintain a "strong military presence" in the region. Given the obvious general thrust of the new Administration's national security policies, however, it is not likely that either the Japanese or the Chinese put a great deal of stock in those assurances.

The recent public airing of the so-called "swing strategy" probably added to skepticism throughout the region. The U.S. "swing strategy" and its possible consequences are of sufficient importance to deserve detailed treatment later in this paper. Suffice it for the moment to observe that the heart of the strategy is a plan to withdraw major American military forces—primarily naval—from the Pacific Ocean area and "swing" them to the Atlantic in the event of a Warsaw Pact attack on NATO. In light of the foregoing factors, one can expect both countries, as well as others, to keep a close eye on the size and operational patterns of the Seventh Fleet, watching for any signs that a drawdown is taking place.

Had the 1950 invasion of South Korea succeeded, Soviet warships and aircraft would surely now be operating from bases at the southern end of the Peninsula, dominating the Strait of Tsushima and the Sea of Japan. That they are not was due to swift and resolute action on the part of the United States to defeat North Korean aggression. Today, these ports are closed to the Soviets, and the Peking-Moscow rift ensures that no ports in China will be available, either.

To the south, however, the story is considerably different. Following the American withdrawal from Vietnam, age-old enmities between that country and its giant neighbor to the north surfaced once more. Beholden to Moscow for huge amounts of military aid provided to North Vietnam during the latter's conflict with South Vietnam and the United States, Hanoi has been more than willing to permit Soviet warships to use its ports and naval bases that were developed by the United States. These ships now call routinely at such places as DaNang and CamRanh Bay. This presence, if additional signals were needed, serves—among other things—to caution the Chinese that Vietnam is a client state of the Soviet Union. Further than that, it provides Moscow with a maritime foothold in the South China Sea region, something the Soviets have never before enjoyed. The consequences of this forward deployment of the Red Navy are difficult to overestimate, most especially with respect to the political dividends flowing from a visible naval presence in the area. Moreover, as has already been noted, it is but a short voyage from these Vietnamese ports to Singapore and the Strait of Malacca.

In this regard, it should be remembered that in the immediate post-World War II days, Moscow was known to be fueling the communist insurgent movement in Malaya. After a long and difficult struggle, Great Britain managed to suppress that rebellion, but it was a near thing. This is not to say that Soviet actions in this instance—any more than with earlier Russian efforts in Indonesia and Iran—were an integral part of the current drive for maritime dominance. At that time, the Soviet Fleet was far too small and too weak to have taken advantage of any such gains. The situation today is considerably different, and the Soviet Navy's presence in Vietnamese waters must be factored into the expanding body of evidence which increasingly validates the contention that the leaders in the Kremlin are exploiting the growing capability of the Soviet Navy in a concerted drive for global maritime control.

Across the South China Sea in the Philippines, the Soviet-backed Huk (communist) rebellion appears to be moribund, recent attention focusing almost exclusively on continuing Moslem unrest in the southern portions of the archipelago. So long as major elements of the Seventh Fleet remain based in the Philippines, and the Moslem uprising does not get out of hand, it would seem that there is little chance that the Kremlin could score any significant political or military gains in the region. On the contrary, it is far more likely that if the Philippines were to turn anywhere else than to the United States for assistance, it would be to the People's Republic of China.

At least in the Western Pacific, except for Vietnam, it appears that strategic planners in Moscow are looking in vain for openings to exploit, and the region thus seems relatively safe from further Soviet incursions, at least for the moment.

Confronted with this reality and disposing the newly-acquired outreach provided by the Red Navy, the Kremlin has necessarily turned its attention eastward to the islands of the South Pacific. As the flood tide of nationalism continues to transform colonial holdings into infant, independent nations, each inevitably plagued by economic and other problems, opportunity has again attracted the roving, imperial Soviet eye. Samoa and Tonga, for example, have already heard the siren call issuing from Moscow.

As with Mauritius in the southern Indian Ocean, the Soviet Union used its gigantic fishing fleet in Samoa and Tonga as the opening wedge. First, Moscow offered to provide the islands with attractively priced fish—caught and canned by Soviet ships—to ease the islands' immediate problems of food supplies. Close on the heels of that ploy came the offer to construct modern jet airports and up-to-date harbor facilities. The previous experience in Mauritius makes it reasonably clear that the next Soviet step will be to arrange for rest and rehabilitation calls for the crews of the ships providing the fish. Shortly thereafter, Russian men-o'-war can be expected to make an appearance, as the islands fall more deeply in debt to Moscow. A foothold would thus be established as it has been elsewhere.

More importantly, Soviet warships—and perhaps aircraft—would then sit astride a major Pacific Ocean trade artery. This particular one stretches from the west coast of the United States and the Panama Canal directly to America's main allies in the southwestern Pacific: Australia and New Zealand. The importance of this artery understandably accounts for the intense concern over such Soviet moves expressed in Canberra, Australia, during meetings there in December of 1976 with the commanders of the navies of these two allies.

At that time, these leaders had a good deal to be concerned about. Faced with austere naval budgets in the two countries and worried about the Soviet initiatives they were following closely, both commanders were deeply worried about the American President-elect's campaign pledge to demilitarize the Indian Ocean. Knowing full well that any such Soviet-American treaty would surely permit continued transits through the Indian Ocean by Red warships—under the guise of interfleet transfers between the Atlantic and Pacific—and acutely aware of the immense capability inherent in the other facets of Soviet maritime power, Australian and New Zealander alike were patently beginning to feel increasingly isolated.

Initial concern turned to genuine alarm in early 1977 when the new President, Jimmy Carter—in the wake of Secretary of State Cyrus Vance's abject failure to sell the much heralded American strategic arms limitations proposals in Moscow—proposed immediate Soviet-American negotiations aimed at demilitarizing the Indian Ocean. The fact that neither of these long-time American allies had even been consulted on an issue of such grave importance to them and their security added a large measure of anger to their reactions, and the Soviet blandishments to Samoa and Tonga took on even more ominous overtones.

The present political leaders of Samoa and Tonga did not look with much favor on the proposals put forward by Moscow. Indeed, the Soviet offers were summarily rejected. Nevertheless, it cannot be assumed that the last word on the subject has been spoken. As with the Maldives, there is little assurance that the present leaders of these islands—backed into an economic and political corner by the myriad problems besetting these embryonic nations—will not change their minds, or that some succeeding and less suspicious regime will view the Soviet proposals in the same light.

As the winds of independence continue to blow throughout the Central and South Pacific, it can be expected that further, similar Soviet efforts will materialize. The United States and its allies will be well advised to keep a weather eye on this remote part of the world for any early indications that Moscow has selected yet another target for exploitation.

In the Pacific Ocean, north of the equator, American bases located in Alaska and the Hawaiian Islands confer almost complete maritime domination on the United States. The great circle shipping routes crossing these stormy seas are all within range of U.S. naval and air power. Moreover, there are no shipping choke points in the accepted sense of the term. Nonetheless, the growing reach of the Soviet Navy—even operating from bases such as Petropavlovsk on the Kamchatka Peninsula—poses a threat to these sea lines of communication that cannot be ignored.

One of the foremost such threats centers on American oil dependence. In the wake of the 1973 Arab-Israeli war, the Arab Organization of Petroleum Exporting Countries (AOPEC) clamped a complete embargo on the United States. As a result of the ensuing oil shortage in America and new discoveries on the north slope of Alaska, a new source of U.S.-controlled oil was brought on line. Since it is located in the 49th state, North Slope oil can only reach the "lower 48" by sea via the Gulf of Alaska. Thus, the United States now has a new and vital energy lifeline that is susceptible to Soviet interdiction in the event of hostilities.

In general, eastward from the International Dateline the Pacific Ocean presents to the mariner an almost unbroken expanse of salt water until he reaches the shores of the Western Hemisphere. Thus, there is little opportunity in this area for Moscow to exploit. Once across the vast Pacific, however, opportunity again beckons, and the Kremlin has not missed its signal. Furthermore, recent political changes portend dangers that cannot be ignored.

Where the west coast of South America bulges into the Pacific Ocean, Peru—far larger than its Ecuadorian neighbor—commands the waters south of Panama. The United States recognized the strategic importance of Peru in the opening days of the nineteenth century and accordingly selected Callao, the port serving the Peruvian capital at Lima, as the de facto headquarters station for its tiny Pacific naval squadron. For several decades, until the westward advance of Americans reached and acquired California, sailing ships of the United States Navy, operating from Callao, patrolled the length of the two continents protecting the interests of the nation. Without U.S. access to centrally-located Callao, it is altogether likely that the history and the political shape of the lands washed by the waters of the eastern Pacific would have evolved in a quite different fashion.

Moscow began its own, modern-day campaign in Peru by wooing the army faction of the ruling military government. Soviet tanks and artillery, accompanied by the usual covey of Russian advisers, technicians and support personnel entered the country first. The Peruvian Navy and Air Force, however, continued to look to the north for help from the United States. Following American refusal to sell relatively modern combat aircraft to the South American country, however, Lima turned to the Soviet Union once more. This time, the influx of Soviet personnel was accompanied by SU-22 fighter-bombers for the Peruvian Air Force. Still, the Peruvian Navy held out.

Having long harbored a healthy distrust of the USSR, leaders of the Peruvian Navy also possessed a strong desire to emulate what they saw as the premier naval power in the world: the United States. In terms of the sophistication and capability of modern warships, Peruvian naval requests were exceedingly modest. For the most part confined to World War II-vintage destroyers and submarines, with armament involving nothing more sophisticated than the post-war ASROC (a rocket-thrown antisubmarine weapon), the request could not be considered out of line with naval weaponry the United States was transferring to other nations. But two political considerations intervened to thwart attempts to meet Peruvian desires.

Within the Pentagon, the presence of Soviet aid programs and Russian military personnel in the country was bound to create objections. On Capitol Hill, a long-standing anger over periodic Peruvian seizure of U.S. fishing boats and the concern voiced by conservatives about the friendly relations between Peru on the one hand and Cuba and Panama on the other made it extremely difficult to obtain the requisite Congressional approval for the transfer. In the event, Peru succeeded in acquiring only a small fraction of the naval weapons systems it sought, and a reluctant Peruvian Navy ultimately also turned to the Soviet Union. One can expect that the military support picture in Peru will now be rounded out with the arrival of still more Soviet personnel, this time accompanied by modern missile-firing boats which will join the American and British warships already in Peru's naval inventory. Not to be overlooked is the fact that with this American refusal, any chance of influencing the future policies of Peru—through the heretofore staunchly pro-American Peruvian Navy—has been lost.

Can one soon expect to see Soviet warships routinely calling at Callao? Will the shift in Peruvian orientation and the recently negotiated U.S.-Panamanian canal treaty result in a heretofore unprecedented Soviet naval presence in the eastern Pacific Ocean? These questions highlight one of the fundamental weaknesses of U.S. foreign and military policy-making: compartmentalization.

No reasoning person can today dispute the fact that all who inhabit this planet live in an interdependent world. Events happening ten thousand miles away, no matter how unimportant they may seem, can nevertheless have profound effects on the national security or well-being of another nation—regardless of the power that nation possesses. It follows that when national strategy planners sit down at their desks and attempt to peer into the future, they must comprehend and take into account the ripples—sometimes the tidal waves—which planned actions of their own or the perceived actions of others are likely to generate. The first of these two considerations—the probable effects of one's own actions—is the one most often ignored by U.S. policymakers. The Panama Canal provides a classic case in point.

It can be fairly inferred that the foremost—if not the sole—factors governing the U.S. Government's approach to the problem of Panama were elimination of the "Big Brother" image which has plagued American relations with the nations of South America for more than a century; the anti-colonialism facet of the modern American ethic; and the altogether practical desire to get some of the Latin American nations, the United Nations, and the bulk of the so-called Third World "off the American political back." Several American Presidents apparently saw the problem in these terms. It is not at all clear, however, that any of their administrations made a serious effort to place the projected solution in the context of either regional or global U.S. strategic interests; nor does it seem that adequate thought was accorded the ramifications that the proposed action would have in the future. Without this sort of analysis, the decision to accept the kinds of treaties ultimately negotiated can only be characterized as narrowly focused, tactical efforts to solve immediate problems without proper regard for the future effects of such actions.

One wonders why the evident transformation of the strategic maritime situation in the Caribbean Sea and the eastern Pacific Ocean—a change which would inevitably accompany Soviet access to a Panama Canal controlled by a nation friendly to Cuba and the USSR—did not receive more attention. By this, one means not the denial of access to the United States—something so obvious that it was debated endlessly—but rather the affording of similar entree to the Soviet Union. To be sure, such use is available to the USSR today, but has not been exploited by the Soviets for understandable reasons: the well-known Soviet paranoia over naval security, reluctance even tacitly to admit dependence upon an American-controlled waterway, and others.

Once the stewardship of the Canal is shifted to the Panamanians, such Soviet inhibitions are likely to disappear, just as they did with respect to transits of the Suez Canal. When that happens, the USSR will have full and easy access from Cuban ports to the eastern Pacific via the Canal, something it does not presently enjoy because of reluctance to utilize an American-controlled waterway and the great distances involved in transits from Soviet Pacific bases.

Will the world soon be witness to Soviet warships calling routinely at Callao in Peru and the materialization of some sort of permanent Soviet naval presence in the eastern Pacific Ocean? So long as the United States continues to operate the Canal, and in light of the great distances across the Pacific, such a development does not appear to be likely. This is not to say that infrequent visits in the near term can be completely ruled out. And as the political status of the Panama Canal inevitably shifts in the future—as a result of the recent treaties—it can be assumed that Soviet naval transits of that waterway will eventually occur and will then gradually increase in frequency, just as they have in the case of the Suez Canal.

Once the Panama transformation begins, Callao will become easily accessible to Russian ships from the Atlantic, probably calling at Havana or Cienfuegos enroute. How soon Moscow will take advantage of the Canal's evolving political character remains to be seen. As already suggested, warships themselves cannot be expected to lead the way; they will come later. Other elements of Soviet maritime power will appear at first: oceanographic survey vessels, the ubiquitous fishing craft, perhaps even space-event support ships—the latter often performing naval functions, including periodic service as flagships for Soviet naval commanders. Based on Moscow's previous actions around the rest of the maritime world, prudent American strategic planners will have to watch developments in this crucial region very carefully.

6. Implications of the Soviet Campaign

With a west-to-east transit of the Panama Canal, this brief survey of global Soviet politico-maritime machinations returns to its starting point in the Caribbean Sea. To the careful observer, the pattern seems abundantly clear, and the ultimate objectives driving the effort are not difficult to infer.

One must conclude that the indispensability of meaningful sea power to any nation aspiring to international influence or dominion has at last been recognized by the Soviet leadership. The spectacular rise of the Soviet Navy, the unprecedented expansion of the Russian oceanographic fleet, the continued growth and modernization of the merchant and fishing fleets—principally within the last two decades—and the manner in which all of these forces are being employed today leave an observer with no other rational explanation. Moreover, it is next to impossible to escape the belief that the importance of shipping choke points in an interdependent world, the overwhelming proportion of whose trade must travel in ships sailing across Mahan's "broad common," has penetrated the Russian consciousness as well.

Whether the Soviets have been studying the history of the British Empire and the writings of Alfred Thayer Mahan is not terribly important. What is significant, however, is the evidence pointing rather clearly to the fact that in recent years the Kremlin has moved decisively to exploit the lessons that can be derived therefrom; and that Moscow is currently engaged in a concerted drive for maritime superiority—particularly for control of key sea lanes which crisscross the seas and, most especially, for dominion over the choke points through which the bulk of the planet's shipping passes each day. Furthermore, one would be naive in the extreme to believe that the challenge thus mounted is receding rather than growing—growing not only in size but in sophistication as well.

From the American point of view, the Soviet challenge is troubling, first of all, because the United States is not now and probably never has been self-sufficient in those natural resources which are indispensable to a functioning and growing society. This fact has never been more pertinent than it is today: At no time in its history has the United States been so dependent upon use of the world's sea lanes for its economic and, therefore, political well-being. Thus, any threat to that use must be taken as exceedingly serious. Second, when an attempt is made to devise the best means of countering such a threat, one is confronted with two acute problems: the traditional beliefs and policies of the North Atlantic Alliance; and, of greater import, the national security policies of the Carter Administration.

Since the inception of the North Atlantic Alliance, that body's member-states have clung tenaciously to the precept that the defense perimeter of the Alliance which encompasses its security interests extends landward only to the borders of the allied nations themselves. At sea, that perimeter is taken to include the waters of the Mediterranean Sea and those of the Atlantic north of the Tropic of Cancer. It should be noted that the only reason this Atlantic boundary is not considerably farther north—the latitude of Gibraltar, for instance—is that a very large portion of the southeastern continental United States would then be left out of the NATO treaty area. The concept of such a limited defensive maritime perimeter in the modern interdependent world of today is obviously anachronistic. Nevertheless, it is a fact of life in NATO, and suggestions to move this maritime border farther south have traditionally met with adamant refusal on the part of European members of the organization. Only the United Kingdom has been an exception.[3]

The fundamental reasoning underlying this otherwise incomprehensible stand has been relatively straightforward: The Europeans do not want to risk being inadvertently drawn into some foreign conflict deriving from the worldwide commitments of the United States or the United Kingdom. Over the years, as the British Empire dissolved under the pressure of nationalism, the European fear of such an eventuality has come to focus exclusively on actions of the United States. Vietnams and airlifts to Israel, for instance, are anathema to the Europeans, and they want to be sure that they can remain aloof from such endeavors.

The problem, however, is that the NATO nations are today as dependent upon overseas resources and markets as is the United States. In fact, in many categories—for instance, petroleum from the Middle East and minerals from Southern Africa—they are infinitely more dependent. As the 1973 Arab-Israeli war clearly demonstrated, West European industry is actually held hostage by Middle East oil. Should that energy lifeline be severed, by whatever means, the European part of NATO would face immediate and undiluted disaster. Much the same situation obtains with respect to raw materials from Africa and many of the Indian Ocean countries. Yet, only recently have these same NATO nations consented

---

[3]France does not subscribe to the general European view and maintains large naval forces in the Indian Ocean, for example. But then, France, while remaining a member of the North Atlantic Alliance, left NATO, the military arm, in 1966.

even to discuss the subject of moving the Alliance's maritime boundary south of the Tropic of Cancer.

Quite clearly, their economic well-being and national security rest on the unfettered use of the sea lanes which connect them to the sources and markets they must have to survive. No longer can they expect a free ride on the backs of the greatly diminished United States and Royal Navies. One is constrained to observe that an equitable sharing of the Alliance's maritime burden is just as much a requirement as contributions to defense of the land borders separating NATO from the Warsaw Pact countries. Until this reality is forthrightly addressed and an Alliance solution is not only agreed upon but implemented, the North Atlantic Treaty Organization will continue to be increasingly vulnerable to the expanding Soviet maritime prowess.

The theories of some notwithstanding, the Russians have not built almost 250 attack submarines solely to counter the U.S. sea-launched, ballistic-missile threat to the Soviet homeland. Anyone who does not expect to find the bulk of those boats prowling the globe's oceans in a drive to sever the sea lines of communication connecting North America to Europe, as well as those sea lanes over which critical raw materials flow to Europe and the United States, is suffering from a crippling delusion. Furthermore, the growing threat posed by the sophisticated surface combatants and aircraft of the Red Fleet—operating out of many strategically located facilities around the globe—is equally real.

Setting the Alliance to one side and considering the Soviet challenge strictly from the American point of view, one finds much to be concerned about in the United States. Over the past few years, the U.S. Navy has steadily shrunk from a fleet of more than 900 ships to less than half that number today. Of equal concern is the fact that in the past two and a half decades, the number of overseas bases accessible to U.S. naval and air forces has diminished from 150 to about 30. During this same period, the Soviet drive for maritime superiority has raised to roughly the same level the number of similar facilities available to the USSR. The obvious deleterious effect of this decline in the number of overseas bases on the U.S. Navy's ability to react in a timely manner to a distant crisis, and the accompanying decrease in staying power far from American shores, coupled with the shrunken size of the Fleet itself, render the American naval situation more critical than a cursory examination of the numbers would suggest. The practical effect of these reductions in numbers of U.S. ships is starkly revealed in Graph II. Readily apparent is the precipitate decline in U.S. Navy ship-days beyond home waters which began as the 1970s dawned. Only the ongoing war in Vietnam kept the overall deployment figures from falling below the reduced levels shown in the early 1970s. Once that war was terminated, time spent in the Pacific plummeted, as it had already done in other regions. To understand more fully the import of these changes, one should examine Graph III in which deployments of the U.S. and Soviet Navies are directly compared.

Although confronted with a spectacular expansion of the Soviet Navy and its other forms of maritime power, the United States has nonetheless allowed the U.S. Fleet to decline further to a strength that is now well below that of the Soviet Union. As a matter of fact, the U.S. Navy is today smaller than at any time since the early days of the 1930s. And it is important to realize that the United States no longer has a powerful, wide-ranging Royal Navy at its side to help meet the increasing maritime challenge being mounted by Moscow.

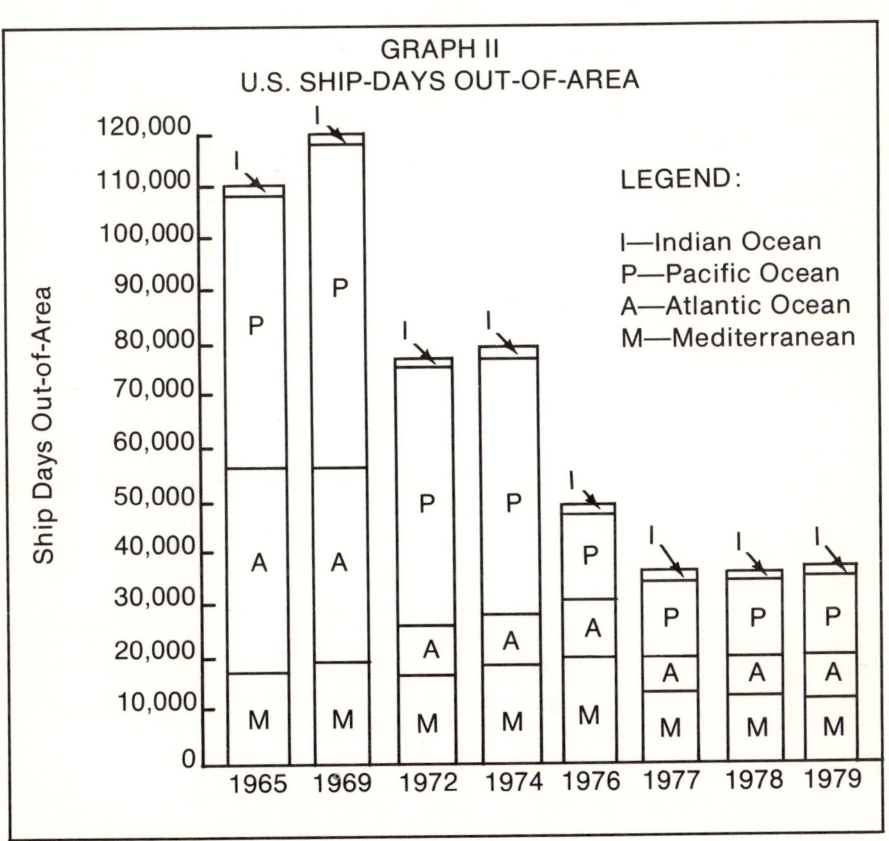

GRAPH III

TRENDS IN U.S.-SOVIET NAVAL OPERATIONS:
SHIP-DAYS OUT-OF-AREA

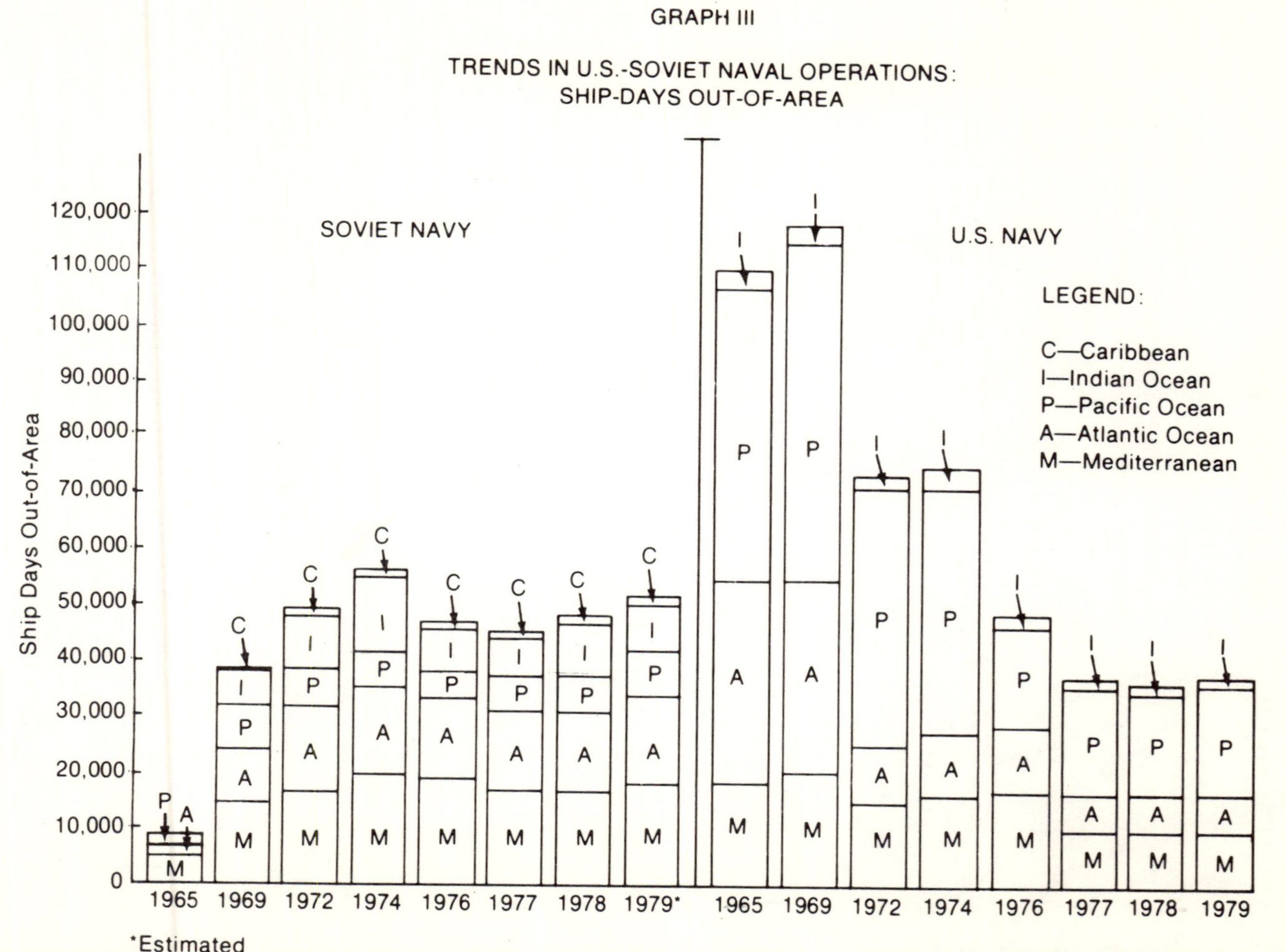

*Estimated

7. U.S. and Soviet Naval Comparison

If it is true—as argued herein—that global maritime competition realistically involves only the navies of the USSR and the United States, it is equally true that attempts to compare U.S. and Soviet naval strength are central to the issue. Yet, such comparisons inevitably lead to vehement arguments over the yardsticks to be used. Whether one selects numbers of ships, their capabilities, their susceptibility to severe damage or disablement, the competence and combat effectiveness of their crews, or whatever, a debate is bound to rage, and the results are most often inconclusive. Nonetheless, there are a few simple verities with which only the most tightly closed mind can argue.

The United States is today an insular nation, critically dependent upon use of the seas to keep the American economy running, to communicate with U.S. allies in the interests of this nation's security, and to permit naval operations beyond the seas during time of war. Given the foregoing factors, the United States clearly must possess the capability to exploit the seas which girdle the globe. This does not, of course, mean that the U.S. Navy must be capable of exercising absolute control over all of the world's oceans at all times. It does mean, however, that America must be able to ensure safe and unfettered use of discrete portions of the oceans, when and where they must be utilized.

As opposed to this requirement—which dictates that the United States should possess a maritime-oriented national strategy—the Soviet Union is far less susceptible to such imperatives, since it is not commensurately dependent on the seas for survival. Moscow's maritime objectives stem, on the other hand, from a widely different set of circumstances: central, continental, geographic position; abundant indigenous natural resources; and an imperial political drive. Thus, the Soviet Union's strategy is diametrically different from that of the United States. Even if we grant dedication of a portion of Soviet naval assets to protection of the Russian sea-based strategic submarines—a proposition rendered exceedingly questionable by the advent of the *Delta*-class ballistic missile submarines with their 4,000-mile-range missiles—two prime maritime objectives must be ascribed to the Soviet Union by prudent U.S. national security planners: first, denial to the United States of the use of the sea lanes for the import of strategic materials, without which the nation's war machine could not long function; and second, isolation of the United States from its overseas allies, along with interdiction of U.S. ability to reinforce and supply those allies.

Hence, the currently popular shorthand for describing the two navies characterizes the United States Fleet as a so-called sea-control force and that of the Soviet Union as a sea-denial force. It is important that these terms and their import be understood. Guerrilla warfare provides an illuminating analogy.

The sea-control mission can be likened to the problems facing a government beset by guerrilla insurgency. It must ensure security of all vital installations, all of the time. The loss of even one power plant, a single oil-storage facility, or a military complex can cause untold havoc. Consequently, extensive guard forces are necessary, forces which spend a good deal of time in unproductive patrolling. It is from this reality that derives the accepted view that a government must dispose a troop ratio of somewhere between five and ten to one if it is to have a reasonable chance of defeating an organized insurgency.

On the other hand, a sea-denial force can be likened to a guerrilla movement, since it possesses the invaluable advantage of the initiative. Free to choose the time and place to strike, the guerrilla leader can plan for and achieve local superiority, wreak damage out of all proportion to the overall number of troops he commands, simultaneously keeping massive numbers of defenders tied down elsewhere. The guerrilla can thus fight an opponent successfully even though that opponent disposes many times the strength of the guerrilla band.

These same principles apply to warfare at sea. Interested essentially in sea denial, the Soviet Union can afford to field a much smaller fleet than can the United States. A half dozen Soviet submarines deployed along the oil transit lanes in the Indian and South Atlantic Oceans, for instance, could demand an immense U.S. or allied effort to counter, in order to ensure the survival of enough tankers to keep adequate supplies of petroleum available at the user's end of the pipeline. Given these factors—which two world wars have abundantly validated—one would expect to find the Soviet Navy, in numbers, running an appreciably distant second to that of the United States.

This situation would obtain even if the United States by itself—or in conjunction with its allies—dominated every significant shipping choke point on the globe. If, however, one factors in the evident Soviet campaign to control those choke points—along with Moscow's gains to date—the required strength quotient for American or allied sea power begins to rise geometrically. For, while some of these maritime funnels can be circumvented or bypassed—albeit at the cost of lost time and increased length of voyages—many cannot. In the latter instances, threats to use of the sea lanes mounted from land bases proximate to the choke points must be neutralized. In general, this means eliminating the bases, either by destroying them or by physically seizing the territory that harbors them and eliminating the threatening forces. Thus, not only are greater numbers of defensive ships required for the so-called sea-control function, but the demand for strike forces such as attack aircraft carriers and Marine amphibious landing units escalates as well.

Thus, yet another verity comes into play. Despite the protestations of those who disparage the need for numbers of ships—in debating the comparative prowess of the U.S. and Soviet Navies—numbers *are* important. As has been pointed out all too frequently, regardless of the greatly increased capability of today's warship, as compared with its counterpart of the past, that man-o'-war cannot be in two places at the same time. If one accepts the foregoing premise, an examination of relative numerical strength becomes relevant, if not central.

A comparison of the active U.S. Fleet of 1979 with that of 1939, and then with the Soviet Navy of today is at once illuminating and sobering.

As revealed by Table I, the American "one ocean" navy of 1939 totaled 373 ships as compared with 446 in 1979. Critics argue that this represents a healthy increase and is really far more than is now needed. While they point to the enhanced capabilities of today's warships as opposed to their 1939 counterparts—and in this they are correct—they overlook the reality that the 446 figure includes 41 fleet ballistic-missile (FBM) submarines and 65 amphibious ships, all performing functions that did not exist in 1939. Moreover, the FBM submarines play no part in discharging the normal responsibilities of modern sea power. Truly comparable figures would, therefore, be 373 in 1939 and 340 in 1979.

Even pro-navalists will readily admit that the ships of the present U.S. Navy are far more sophisticated and capable than were their pre-World War II ancestors. But then, so are those of this nation's potential adversaries. If one therefore considers the threat, and restricts the equation to surface combatants—aircraft carriers, battleships, cruisers and destroyer types—the true cutting edge of naval power, the inescapable conclusion is that, relatively speaking, the U.S. Navy is considerably weaker today (177) than it was in 1939 (227).

TABLE I

COMPARATIVE U.S. NAVAL STRENGTH: 1939 AND 1979[1]

Ship Type	*1939*	*1979*
Aircraft Carriers	5	13
Battleships	15	—
Cruisers	37	27
Destroyers	159	72
Frigates[2]	11	65
Submarines	61	121
Amphibious	—[3]	65
Replenishment & Support	85	83
Totals	373	446
Less Amphibious Forces		65
Less SSBNs		41
Comparative Total		340[4]

Notes:
[1]The distinction between strategic and general purpose forces did not exist in the pre-nuclear era of 1939.
[2]Known as "patrol craft" in 1939; subsequently identified as destroyer-escorts or frigates.
[3] This function did not exist in 1939.
[4]Total with foregoing functions subtracted.

A comparison of the present U.S. fleet with that of the Soviet Union (Table II) reveals that the U.S. Navy is roughly 56 percent the size of the Soviet Navy. Moreover, if one again restricts the examination to those warships constituting the cutting edge of naval power, 177 major American surface combatants confront 275 similar ships wearing the Hammer and Sickle.

Critics of sea power invariably charge that pro-navalists always include hundreds of small patrol craft in their figures to demonstrate overwhelming Soviet superiority. The above totals of 177 and 275 include neither those patrol boats nor the mammoth Soviet submarine fleet. The resulting picture of comparative naval strengths is not exactly a comforting one.

Even less reassuring is the situation when two additional factors are fed into the calculations: amphibious capability, one of the fastest growing segments of the Soviet Navy; and aircraft carrier construction. Moscow currently has two or more carriers under construction (including a 78,-000-ton, nuclear-powered version) while the United States is completing only one—the nuclear-powered *Vinson* which, for almost three years, seemed destined to be the last carrier added to the American Navy.

TABLE II

COMPARATIVE U.S. AND SOVIET GENERAL PURPOSE NAVAL STRENGTH

1979

Ship Type	*U.S.*	*USSR*
Aircraft Carriers	13	4[1]
Cruisers	27	35
Destroyers	72	100
Frigates/Corvettes	65	136
Submarines	80	248
Amphibious	65	85
Replenishment/Support	85	119
Totals	405	727

Note:
[1]Currently, these ships are advertised by the Soviets as anti-submarine ships and do not possess the capabilities of American attack carriers. The *Kiev* and her follow-on sister ships, however, have the potential for other missions—fleet air defense, amphibious operations—and may develop a potential strike capability if the V/STOL technology eventually produces a truly high-performance aircraft. Highly accurate reports in early 1980 credit the USSR with having a 78,000-ton, nuclear-powered carrier under construction. This, of course, will mark emergence of the Soviet Navy with strike capability across the board. The degree to which this development will alter the current naval balance will depend on how many of these ships they ultimately build.

Both countries are building new ships at present. In the United States, the aforementioned *Vinson* is due to be commissioned in 1982. The remainder of the ongoing U.S. effort is concentrated, insofar as true men-o'-war are concerned, in destroyers, frigates and submarines. In this connection, the new *Spruance*-class destroyer program is virtually complete. The *Oliver Hazard Perry*-class of anti-submarine-warfare frigates is well under way at a rate of six commissionings a year and may eventually total 30-40 in number. A follow-on modification of the *Spruance*-design will be produced, and probably a dozen of them will be built over the next five to ten years, each incorporating the advanced Aegis air-defense system. Since the latter ships will be somewhat larger than the original *Spruance* versions (about 9,000 tons) the U.S. Navy has recently decided to designate them cruisers rather than the originally-selected terminology, "guided-missile destroyers."

This action may serve, to some extent, to mask the actual situation regarding cruiser construction in the U.S. Navy. Considerable controversy has attended this class of ship over the past few years. Congress long ago mandated that warships in excess of 8,000 tons must be nuclear-powered, although this stricture will evidently be overlooked in the Aegis-equipped *Spruance*-class ships now programmed. Plans to build more of the *Virginia*-class nuclear-powered cruisers previously encountered heavy weather in the Executive and Legislative branches, and cruisers have subsequently been eliminated altogether from the current five-year U.S. shipbuilding program.

An effort was made by the Navy a few years ago to launch a building program designed to produce a "strike cruiser," which would have been a warship of some 20,000 tons, bristling with guns and missiles, aircraft and anti-submarine weapons. With its unprecedented combat capability, this class of ship was to have been employed as part of the defensive screen of carrier strike groups operating in high threat areas, as well as in independent operations where the ship's defensive capability would permit it to survive on its own. The price tag—around one billion dollars per copy—killed the concept, and the ship never got off the designer's drawing board. Based on this experience, one must conclude that any sort of cruiser program in the U.S. Navy is presently dead.

This situation is in sharp contrast to recent shipbuilding developments in the Soviet Union. Reports have been confirmed that new classes of cruisers not only have proceeded through the design phase but are entering construction. Heading this list of new ships is a huge, 32,000-ton monster being built at Leningrad. The *Kirov*-class of nuclear-powered cruisers appears to be an upgraded version of the now defunct American strike cruiser. Perhaps twelve of these ships are in the works and, given the confirmation that a new attack aircraft carrier is on the way in the Soviet Union, one can assume that the missions of these ships will be similar to those envisioned for the U.S. carriers and the ill-fated American strike cruisers.

Revelation of the foregoing two Soviet naval building programs should be enough to give pause to any Western naval planner—but there is more. In addition to the *Kirov*-class, which will be nuclear-powered, there are at least three other varieties of cruisers in the works. Also nuclear-powered, these additional cruiser classes will reputedly fill anti-submarine, sophisticated anti-air warfare, and cruise missile roles. Moreover, it appears that construction of a new class of large, heavily-armed, logistics ships is also underway.

To round out this picture, we need to look at Soviet naval shipyard capacity. Here again, recurrent reports credit the Soviets with very large capital investments devoted to expanding and modernizing their naval shipyards—a development which, while not necessarily of eye-catching interest, is nevertheless indicative of future intentions.

Thus, by examining the ongoing naval efforts of the two superpowers, it is possible to identify trends having significant import for the future insofar as use of the world's sea lanes is concerned. From the Western point of view in general, and that of the United States in particular, the outlook is anything but comforting.

8. The Swing Strategy

There are other factors that impinge on the true relative strength of competing navies. Among them are the individual and collective competence of ships' companies; effectiveness of damage control, i.e., the capacity of men-o'-war to sustain combat damage and continue to fight; weapons loads; and geographic constraints. All too often, Western anti-navalists argue that numbers of warships really make little difference. Specifically, they assert that the U.S. Navy has been and is far superior to the Soviet Navy, because of the former's advanced technological sophistication and its clear edge in attack aircraft carriers. While such arguments have merit, they do not obviate the necessity for the United States to build and deploy a fleet that is sufficiently numerous to meet probable challenges in vital oceanic areas. The very existence of the swing strategy lends credence to the contentions of those who claim that numbers—despite advantages accruing from technological superiority—are indeed important.

The so-called swing strategy, as previously noted, envisions the transfer of significant U.S. military strength from the Pacific Ocean to the Atlantic theater in the event of a Warsaw Pact attack on NATO. Although some Army and Air Force units would be involved in the shift, the main power subject to the swing would be naval. And the perceived need to shift such naval forces stems from the Administration's implicit judgment that the U.S. Navy does not possess enough ships to permit it to fight a major war in the Atlantic without sizable augmentation, and that the requisite additional strength can be drawn only from the Pacific Fleet.

In late 1979, when the internal governmental debate over the U.S. swing strategy broke into print, by way of a Roland Evans and Robert Novak newspaper column, many media commentators traced the concept's origin to a transformation of U.S. national strategy engineered in 1970 by President Nixon. At that time, it was decided that America's goal should be to fashion and maintain sufficient military prowess to fight a full-scale war—either in the Atlantic or the Pacific region—while reserving sufficient capability to deal with a serious international crisis elsewhere. This "1½ war" strategy replaced the "2½ war" variety which had come down from previous Administrations. In the latter case, the advertised aim of the United States was to be ready to fight the Soviets in the Atlantic, the Chinese in the Far East, and still have enough resources to deal with a lesser contingency somewhere else in the world.

The fact of the matter was that, while U.S. national strategy might have held the "2½ war" capability to be the aim, the United States had not boasted the means to achieve such an objective since the days just before demobilization began to diminish American military power at the end of the Second World War. In recognition of that fact, U.S. national security planners, well before the Nixon period, set to work devising plans which, in reality, proved to be the genesis of the present swing strategy. Thus, the notion, which underpins it, has been around for a long time. For eminently practical political reasons, however, the existence of this American plan has never been officially acknowledged by the U.S. Government.

The early 1960s brought John F. Kennedy to the presidency and Robert S. McNamara to the Defense Department. Along with the latter came a group of civilian analysts programmed to apply to the American defense establishment the same methodology that had produced great success at the Ford Motor Company. This new group immediately set about its assigned task of making the American armed services "cost effective," and retrenchment was the name of the game as they sought ways of effecting economies in the nation's defense budget. One of the first targets of the new campaign proved to be the United States Marine Corps, along with the amphibious shipping capable of moving it to trouble spots around the world. With two Marine divisions positioned on each coast of the United States, these analysts reasoned, the nation could get by safely with sufficient shipping to lift only half of that force—two divisions—at a time.

For example, if war were to erupt in Europe, the amphibious ships normally stationed in the Pacific could be "swung" through the Panama Canal to the East Coast of the United States where they would join their Atlantic counterparts, and the combined lift could then transport the two East Coast Marine divisions to where they were needed. Having done that, the entire fleet could then sail back across the Atlantic, through the Canal, to ports on the opposite U.S. coast where it would then be ready to move the remaining two divisions to some other trouble spot, if necessary.

The motivation behind such a plan was relatively simple. By this means, the number of amphibious ships in the active fleet could be cut in half, and so could future construction programs for these expensive craft. Budgetary savings—near and long term—would be sizable, and the cost-effectiveness aim would be achieved. Having made this initial scheme stick, the analysts turned their attention to other elements of the armed forces where they might apply the same methodology, with similar results.

At the time, considerable pressure was being exerted by Senator Mike Mansfield to reduce American military strength in Europe. The amphibious-lift formula—essentially amounting to dual tasking—appeared to McNamara's analysts to offer a way to meet the thrust of Mansfield's demands and, at the same time, effect still further economies in the U.S. defense budget. In this instance, the notion was to withdraw from Europe two brigades of the 24th Division—then stationed in Germany—along with commensurate Air Force strength, and "re-deploy" the entire contingent to the United States. The idea was broached in NATO circles and subsequently implemented.

The troops and aircraft were to be brought back to the United States, stationed at continental U.S. Army and Air Force bases, and our European allies were to be assured that, should the Warsaw Pact threaten an attack upon the North Atlantic Alliance nations, these "dual-based" forces—and more—would be immediately returned to European soil. Implicit in this American pledge—made to elicit NATO's agreement to the plan—was a U.S. commitment to respond with its full military power against any threat posed to any or all of the NATO nations.

To prove to NATO that the U.S. commitment was not thereby being reduced, Washington promised to conduct periodic exercises during which U.S. forces would be flown back to Germany and joined with their prepositioned equipment and supplies, thus demonstrating the feasibility of the plan. In the event, NATO agreed to the proposal—albeit reluctantly—and the American forces came home. Unfortunately for the North Atlantic Treaty Organization, the advertised American plan was almost immediately exposed as something of a fraud. For when the Soviet armed forces invaded Czechoslovakia in the summer of 1968, it was suddenly discovered that the "dual-based" forces had been drastically reduced to meet American commitments in Vietnam. Had the Europeans harbored any initial doubts about Washington's true intentions, those doubts were now confirmed.

It is clear from the foregoing examples that the notion underlying the swing strategy is not new and, certainly, did not originate with the Nixon Doctrine. The attempt to produce economies in the U.S. defense effort by this strategem of dual tasking persists today and has its foremost current manifestation in the swing strategy. This is not to say that, if American national security can be maintained at an acceptable level with such a strategy, it should not be utilized; most assuredly it should be adopted, given the skyrocketing cost of modern weaponry, especially ships and aircraft. On the other hand, if such a scheme will further degrade an already weakened defense posture—and recent Soviet actions in South Asia suggest that Moscow regards both American will and American military strength as too feeble to counter Soviet international adventures—then options such as the swing strategy should be abandoned. To judge the efficacy of the idea, one must examine its possible ramifications.

To begin with, there is the question of how U.S. military leaders evaluate the U.S. international position, weighing threat against capability. One can safely assert that, from the outset of the McNamara era in defense planning, the vast majority of these leaders have looked with considerable alarm on the cost-effectiveness drive and the way in which it was being implemented. In their judgment, too many of the new underlying assumptions were premised on an exceedingly naive view of the international milieu rather than on the hard, cold facts of global life and power politics. They believed that the postulated environment bore little resemblance to reality and that national security planning based on such a rationale was essentially a prescription for ultimate disaster. The recent enlightenment of President Carter, which the Soviet action in Afghanistan produced, has undoubtedly generated a good deal of relief within the American military hierarchy.

Indicators that a relative decline in U.S. armed strength was reaching perilous levels have been evident for a considerable period of time. With respect to the swing strategy, the Commander-in-Chief of all U.S. military forces in the Pacific Ocean area made his position brutally clear when queried during a review of national strategy undertaken in 1979 to develop the Secretary of Defense's so-called consolidated budget guidance to be given to the Joint Chiefs of Staff. Surveying the situation in the Pacific, that commander advised Washington that, at present, he was only "marginally" capable of performing his tasks with his currently assigned forces; that if any significant fraction of them were to be withdrawn for use in the Atlantic, he would have no alternative but to adopt a strictly defensive posture to survive. He added that this would mean abandonment of all his responsibilities in the Western Pacific.

This shocking assessment apparently had little effect on those conducting the study. Yet, it is a startling statement. Presumably, what he was telling Washington was that if the swing strategy were implemented, he would have to pull all his forces back to a line running from Alaska to Hawaii and attempt to defend from there. The incredible picture of a seriously depleted U.S. Pacific Fleet retreating to the Hawaiian Islands and Alaska—thereby placing these two states in the very forefront of the battle area—should, by itself, have been sufficiently unsettling to those involved in the study. The further prospect of having to abandon commitments to all U.S. friends and allies in the Far East, leaving them to contend with the huge Soviet land, air and naval forces deployed to the region should have been enough to call the strategy into serious question. Based on reports in the press, however, it did not. The only concession the analysts and policy-level officials made to these projections was to recommend to Secretary Harold Brown that the existence of the strategy continue to be withheld from American friends and allies in the Western Pacific.

While not exactly commendable, that recommendation is eminently understandable. If the swing strategy is, indeed, to remain a fundamental part of U.S. national security planning, it would do serious damage to U.S. relations with those states if the American Government were to acknowledge publicly a plan to leave them in the lurch should war break out. One must assume, of course, that these same nations have long been privately aware of the plan—and considerably unsettled by its implications. Knowledge of the swing strategy no doubt played a significant part in generating vociferous Far Eastern objections to the Carter announcement, at the outset of his term in office, that he intended to remove all U.S. ground forces from South Korea as quickly as possible.

Given the serious unraveling which has occurred in the northwest quadrant of the Indian Ocean in the past few years—and the accompanying, some would charge contributing, failure of U.S. foreign policies toward the region—there is cause for concern that the reasonably stable conditions prevailing in the Far East today may be similarly undermined by the swing strategy. For example, abandonment of the Western Pacific by U.S. military forces would leave Japan dangerously exposed and relatively defenseless. Because of restrictions in its postwar constitution (dictated by the United States) that prohibit Japan from acquiring the armed strength needed for self-protection, the Japanese have been forced to rely on America for the foundation of their national security. Thus, withdrawal of the U.S. military umbrella could very likely push Tokyo into adopting a neutral stance in the face of any Soviet threat or perhaps to seek some form of accommodation with Moscow.

The subsequent reaction of the People's Republic of China could produce even more profound changes with respect to the strategic situation—not only in the Far East but in Western Europe as well. It is not difficult to visualize a China, flanked by a neutral or Soviet-dominated Japan, deciding that its own best national interest lay in promoting rapprochement with Moscow. The impact of such a massive shift in the balance of global power would be hard to underestimate.

It should be noted that over 40 Soviet divisions and a comparable number of combat aircraft squadrons—along with a sizable portion of the Soviet fleet—are deployed in the Soviet Far East. Because these ground and air forces, in particular, guard the Soviet Far Eastern flank, they are not available for deployment to the European front—a factor that contributes to the security of the North Atlantic Alliance. Thus, many strategists consider China to be a de facto member of NATO for this very reason.

There is yet another factor that must not be overlooked. During the Second World War the amount of military aid provided to China was minuscule compared with that sent to Western Europe and the Soviet Union. What aid did manage to make its way into the country traveled a tortuous route through Southeast Asia and then laboriously by air over the famous "Hump." The reason for this arduous process with its limited results was the presence of the Japanese Navy, which effectively cordoned China off, isolating the entire nation from direct contact with its allies. A present-day retreat on the part of the U.S. Navy would place the Soviet Far Eastern fleet in a similar commanding position, with clearly predictable effects on the political outlook and military calculations in Peking.

The ramifications of the swing strategy do not, however, end here. Surely, one of the first results would be the unification of Korea—on Pyongyang's terms. Moreover, one would expect to see Soviet naval domination of the entire South China Sea—with the Soviet Navy operating initially from American-built bases in Vietnam—and this would certainly impact on the Philippines and Indonesia. Finally, there is the even greater danger that would occur as a result of the loss of control in the vitally important Indian Ocean region and its critical arm, the Persian Gulf. The present Chief of Naval Operations, Admiral Thomas Hayward, has described the consequences of American abandonment of the Western Pacific as "incalculable." Perhaps a better characterization would be "catastrophic."

This examination of the swing strategy serves to highlight the predicament in which the United States and the West find themselves as they face the manifest Soviet drive for global maritime hegemony. The very existence of the strategy suggests that those who argue that numbers of ships are not terribly important and that the current size of the American Navy—given its technological sophistication—is entirely adequate to the nation's needs are dangerously wrong. That existence and the trauma Washington is going through in these dawning days of the 1980s, as it seeks to bolster its position in the Indian Ocean in the wake of the Soviet invasion of Afghanistan, make it clear that the "adequate-number" canard needs to be promptly laid to rest.

9. To Meet the Challenge

The hard fact of the late twentieth century is that America is confronted by a hostile navy, which is superior to its own in overall numbers and in most categories of significant ships, and the trends are decidedly unfavorable. Moreover, these worrisome trends now include categories of naval prowess in which the United States has long held decisive margins of superiority. The fundamental question then becomes: What is the United States planning to do about it?

The best clues to President Carter's defense plans surfaced in Secretary of Defense Harold Brown's preliminary decisions on the naval budgets for fiscal years 1979 and 1980. In both instances he tentatively ordered deep cuts in ship and aircraft development and procurement funds. The clear indication was that, in general, programs would be eliminated, postponed or stretched out in order to pare those budgets. In late September 1977, Navy Department "reclamas" to those initial judgments succeeded in getting funds restored for some programs, but the thrust of Brown's decisions remained.

This was confirmed when Secretary Brown sent the Administration's first five-year, naval-shipbuilding program to Capitol Hill. It called for the following numbers of ships to be constructed: FY 79—15; FY 80—15; FY 81—17; FY 82—12; FY 83—11. The clear implication, confirmed by Administration officials, was that, far from any increase in the size of the Navy over the ensuing five years, its projected strength would continue to fall to the point where the United States would be fortunate to be able to put a fleet of 400 ships to sea by the late 1980s.

It was extremely difficult to understand this trend in light of what was then known about Soviet maritime actions and global initiatives. Of course, there was a very powerful and extremely vocal element within the Office of the Secretary of Defense—particularly among the civilian international security and systems analyst appointees—which seemed prepared to risk the nation's future security on the premise that the only Soviet threat which needs to be met is the one to NATO Europe. Moreover, this group further focused this notion almost exclusively on the European Central Front.

These advocates contended that so long as the United States and its European allies remained strong enough to contain a Warsaw Pact invasion of Central Europe, American and Alliance security would be preserved. The rest of the world and the threats posed to it by the Soviet Union could essentially be ignored.

Further, this entourage of the Secretary of Defense postulated the so-called short-war scenario which assumes that any war in Europe would last less than 30 days. The theory is that before the end of that period, stalemate would result—or the war would go nuclear and be terminated quickly. The problem with this theory is just that: it is only theory. For there has never been a real test of national decisionmaking when nations reach the edge of the nuclear abyss. Some contend, with considerable validity, that such an experience occurred during the Cuban missile crisis in 1962. Since one of the most insane rulers in modern history, Adolf Hitler, did not initiate gas warfare to prevent the utter destruction of his "1000-year Reich"—he had adequate stocks at his disposal—one must admit that it can be argued that other, perhaps more rational, chiefs of state will back away from the dreaded nuclear holocaust when faced with a similar decision. Certainly, the short-war advocates would seem to bear the burden of proof with respect to their untested theories.

The usefulness of their postulation lay in the ammunition it provided for attacking the dollar-intensive naval shipbuilding budgets and, thereby, reducing them. The bureaucratic reasoning was relatively straightforward and uncomplicated. If the war for which the United States prepares itself can be expected to terminate in 30 days or less, navies will not play an important role. Before task forces, convoys or anything else sea-borne can be assembled and brought to bear, the war will be over. Thus, there is no need to spend scarce funds building or maintaining a navy; in fact, it would actually be better to lay up the major portion of the present fleet.

The risks inherent in such reasoning, in the midst of a power-political world, which finds an insular United States becoming increasingly dependent upon overseas sources of supply and markets—not to mention allies—would seem to be manifest. Nevertheless, such policy arguments have commanded considerable sympathy in the Office of the Secretary of Defense as well as in some other agencies of the U.S. Government. As a result, the answer to the question of what the United States was doing about the growing Soviet maritime challenge in the early years of the Carter Administration is startling: the United States was retrenching.

Moreover, events of the past three years, during which the foregoing attitudes persisted, suggest that no one really expected the United States to make any serious effort—either by strong diplomatic measures or by flexing its military muscle—to counter Soviet aggressive adventures or to blunt Moscow's drive for maritime hegemony.

Following the American retreat from Vietnam, waves of apprehension swept through the capitals of many countries allied with or friendly to the

United States. They feared that they would be left to their own devices if the oppressive hand of insurgency—even if accompanied by visible external subversive support—was laid upon their lands. This growing belief that the United States could no longer be counted on to throw its military weight on the side of its friends and allies precipitated a widespread reevaluation of state policies. Then, events in Angola and the Horn of Africa convinced these same nations that neither U.S. military assistance nor political arm-twisting through "linkages" of various sorts could be expected. When the United States demurred at the urging of some to link the Strategic Arms Limitation Talks to Soviet behavior in East Africa, friendly nations could conclude only that, when faced with an international threat to their existence, they would most likely be left to shift for themselves.

Of at least equal importance is the interpretation of those events that was made in the Kremlin. The Soviet planner, sitting at his desk, looking for another shipping choke point ripe for domination, surely must have arrived at similar conclusions: So long as the immediate survival of the United States is not at issue, any sort of serious intervention by Washington is highly unlikely. The portents for further aggressive Soviet incursions and continued prosecution of the shipping choke-point campaign become obvious.

Thus, a second question: What *ought* the United States to be doing to meet this new challenge? Clearly, the nation should be moving on several fronts. First and foremost, the United States Navy should be rejuvenated to the extent that the naval imbalance between the United States and the Soviet Union is, at the very least, redressed. Second, the other facets of sea power should receive more attention. For in crisis situations, how long can this nation prevail when more than 90 percent of its overseas commerce is carried in foreign ships and its welfare is thereby hostage to the political vagaries of these nations? Third, it is also imperative that the full diplomatic and economic assets of the country be enlisted in a coordinated effort to thwart the Soviet design.

Rather than effecting economies in the sea-power strength of the United States, the Administration and the Congress should be looking for ways to re-establish the naval superiority this nation enjoyed at the end of World War II. The most important advantages the U.S. Navy enjoys today over that of the Soviet Union are exemplified by the attack aircraft carrier and the U.S. Marine Corps. So long as manned aircraft remain relevant, the carrier edge must be maintained. But the decline in total numbers of ships must also be reversed. It would serve the nation ill to have a nuclear-powered carrier guarding the approaches to Japan or fighting its way into the Norwegian Sea if the fuel that is indispensable to American tanks in Europe or aircraft defending the Caribbean shipping lanes is being sent to the bottom of the Indian Ocean or the South Atlantic because there are insufficient ships to counteract the Soviet submarine threat.

Whether NATO Europe bestirs itself to safeguard its own interests—something the United States should be vociferously demanding in the Alliance's councils—America cannot afford to neglect its own. If the West Europeans, bemused by their own success while living under the U.S. protective umbrella, refuse to meet the challenge, the United States will have to do the job with concern primarily for its own interests.

We must no longer listen to those currently within the Department of Defense and other U.S. governmental circles who, in their fixation on the land threat to Western Europe and exhibiting a depressing failure to understand the maritime imperatives upon which the future of this insular nation rests, are ready to abandon the oceans to the Soviet Union. The United States ought to be moving in precisely the opposite direction—to win back control of the oceans.

American planners—diplomatic, military, strategic—should be studying this nation's critical dependence on overseas trade and the present Russian initiatives which bid fair to place that trade in jeopardy at any Soviet whim, and should be devising plans to forestall achievement of Moscow's obvious objectives.

In addition to reconstitution of the U.S. Navy, these plans must include the employment of diplomatic and economic efforts as well. In a power-political world, the linking of SALT, for example, to Soviet behavior in various parts of the world makes common sense. The present Administration's adamant refusal to do so served only to convince Soviet leaders that they could have their much sought-after treaty, while continuing to pursue their adventuristic aims around the globe.

The U.S. reaction to the Soviet invasion of Afghanistan highlights yet another consideration—with respect to the SALT II treaty. One of the inherent dangers in the nuclear age is miscalculation. The Administration's rejection of linkage between the pact and Soviet behavior in other areas may have led Moscow to miscalculate the extent to which it could push its imperial designs without endangering the treaty itself. At least for the moment, of course, SALT II is moribund, although Administration spokesmen are still urging its ratification. The miscalculation in this instance is not necessarily critical, since we do not know whether Afghanistan figured more importantly in Kremlin risk assessments than did the treaty. It is cited merely as an example of the danger that U.S. failure to transmit clear, unambiguous signals can lead to mistaken Soviet assessments in far more dangerous circumstances—those involving use of nuclear weapons, for instance.

Termination of industrial exports to the Soviet Union in response to continued Soviet thrusts inimical to U.S. interests and international stability would also be a sensible policy. In this regard, the United States has approved a series of high-technology transfers to the Soviet Union over the past three years that many observers believe to have been unwise. Moscow has been especially eager to procure advanced computer and manufacturing technology, for example. Other political and economic measures can be employed directly against the USSR, and in support of regimes controlling a choke point Moscow has targeted.

As events in Malta and Iceland have demonstrated, the Soviet thrust is not limited to acquiring support facilities in various places, but also encompasses removal of American footholds where they exist, whether or not there is any chance for Moscow to move into the vacuum thus created. There is, for example, a concerted drive underway to evict the United States from its long-held base at Guantanamo Bay, Cuba. Agitation from Cuba, supported by the Panamanian Government, is likely to escalate. Having undertaken to relinquish full ownership and control of the Panama Canal, it would be exceedingly shortsighted for the United States to vacate the naval base at Guantanamo. The strategic importance of that installation has been enhanced by Soviet access to the Cuban ports of Havana and Cienfuegos and by Moscow's ongoing effort to construct extensive naval facilities at the latter location. Should the American Government nonetheless eventually find pressures to withdraw irresistible, the United States should agree only on the condition that the Cuban Government expel Soviet military forces from that island—including the Red Navy from Cienfuegos and Havana—and provide iron-clad guarantees that there will be no return.

In light of the Indonesian experience, it is important that U.S. officials not ignore the political benefits which can be derived from the military training of foreign nationals in the United States. We would do well to remember that, in 1965, when Sukarno and the Indonesian Communist Party attempted to purge the leadership of the nation's armed forces and deliver the government completely into Marxist hands, the leadership of the Indonesian military—the bulk of which had received its training in America—came down on the side of the Free World. The investment made by the United States in such training is often paltry in relation to the ultimate return. Similar training programs should be undertaken as part of a coordinated marshaling of the total force America disposes—political, economic and military—to promote the aims and objectives of the United States and to frustrate those of the Soviet Union.

The Soviet invasion of Afghanistan has apparently shocked the Carter Administration out of its misplaced euphoria with respect to Moscow's true intentions—cloaked as they have been by its opportunistic championing of detente. After three years in office and on the heels of the invasion, President Carter confessed to the nation that, all along, he has been badly misreading the Kremlin's words, deeds and intentions. Subsequently, he delivered perhaps his strongest speech on U.S. national security and the international situation. It almost seems as if the Soviets periodically develop an uncontrollable urge to shore up the resolve of Western peoples to defend themselves and protect their vital interests. Moscow did this previously with a blockade of Berlin, invasions of Hungary and Czechoslovakia, and an attempt to emplace strategic nuclear missiles in Cuba. Now it appears they have done it again with the invasion of Afghanistan. Whether this latest lesson endures, however, remains to be seen.

Early in 1980, President Carter cut off grain shipments and transfers of high technology to the USSR, drew a line around the Persian Gulf and the Arabian Peninsula, and drastically revised his views with respect to defense spending. Whether this attitude will persist is another question. He almost immediately backtracked on defense spending. Many observers are predicting that after a period of time, the Soviets—as they seem invariably to have done in the past—will manage to obfuscate the true aggressiveness of their actions in Afghanistan and mount yet another of their well-known "peace offensives." If the West in general and the United States in particular succumb once more to such Soviet blandishments, the limited measures taken to strengthen American foreign policies and national security are likely to be discarded in the all-too-familiar pattern of the recent past.

President Carter's chief handicap in the present crisis is that, under his direction, the U.S. armed forces have been allowed to continue their deterioration to the point that he is in serious danger of having his blunt words ignored by the men in the Kremlin. For the nation does not have the long-range, quick-reaction force about which so much has been heard lately. Nor is there any assurance that this scheme—which harks back to the Fast Deployment Logistics (FDL) notions of the McNamara era—will work, especially in view of the progressive loss of base and overflight-refueling rights which the United States has suffered since then. The latest concept is still little more than a paper-planning exercise. Moreover, in order to put relevant naval strength into the Indian Ocean, he has been forced to strip ships from the already over-extended U.S. Sixth and Seventh Fleets. The problem is especially acute with regard to aircraft carriers. Interestingly enough, these are the same ships that the President and his appointed advisers had judged to be far too expensive and vulnerable to be useful and whose construction he has consistently vetoed since taking office.

Credible military force depends on a wide variety of modern, sophisticated hardware, as well as on adequate numbers of trained personnel. The lead-times involved in producing both are considerable. This, apparently, is a lesson that the previously cited analysts and policy-level officials have yet to learn. Nowhere is this more true than in the task of fashioning meaningful maritime power. One can only hope that the hour is not already too late.

All of the national security debates currently occupying center stage in the United States will amount to so much worthless rhetoric if the hard choices are not made, and made quickly. Whatever else the USSR is doing—in Afghanistan and Iran, for instance—it seems obvious that it has adopted and is continuing to pursue the goal of global maritime hegemony. The Soviet maritime campaign is not likely to be blunted merely by the words of an awakened President of the United States, especially if he does not command sufficient maritime muscle to give effect to his warnings.

One is constrained to observe that it would be the ultimate irony of history if American international security and domestic economic well-being were permitted to fall victim to a predatory enemy because he was able to best us at sea. Such an end would be supremely tragic if it were to be accomplished by a nation with a long history of land-bound thinking, a nation which only in recent years has learned what the American people, with their three-century, sea-going heritage have known from the beginning: the indispensability of adequate maritime power to an insular nation.

INSTITUTE FOR FOREIGN POLICY ANALYSIS, INC.
List of Publications

Special Reports

The Cruise Missile: Bargaining Chip or Defense Bargain? By Robert L. Pfaltzgraff, Jr., and Jacquelyn K. Davis. January 1977. x, 53pp. $3.00.

Eurocommunism and the Atlantic Alliance. By James E. Dougherty and Diane K. Pfaltzgraff. January 1977. xiv, 66pp. $3.00.

The Neutron Bomb: Political, Technological and Military Issues. By S. T. Cohen. November 1978. xii, 95pp. $6.50.

SALT II and U.S.-Soviet Strategic Forces. By Jacquelyn K. Davis, Patrick J. Friel and Robert L. Pfaltzgraff, Jr. June 1979. xii, 51pp. $5.00.

The Emerging Strategic Environment: Implications for Ballistic Missile Defense. By Leon Gouré, William G. Hyland and Colin S. Gray. December 1979. xi, 75pp. $6.50.

The Soviet Union and Ballistic Missile Defense. By Jacquelyn K. Davis, Uri Ra'anan, Robert L. Pfaltzgraff, Jr., Michael J. Deane and John M. Collins. March 1980. xi, 71 pp. $6.50.

Energy Issues and Alliance Relationships: The United States, Western Europe and Japan. By Robert L. Pfaltzgraff, Jr. April 1980. xii, 71 pp. $6.50.

U.S. Strategic-Nuclear Policy and Ballistic Missile Defense: The 1980s and Beyond. By William S. Schneider, Jr., Donald G. Brennan, William A. Davis, Jr., and Hans Rühle. April 1980. xiii, 61pp. $6.50.

Foreign Policy Reports

The papers in this series of Foreign Policy Reports are addressed to a variety of topics in the field of world affairs, including diplomacy, economics, strategy, science and technology, arms control, international organization, and country and regional issues.

Defense Technology and the Atlantic Alliance: Competition or Collaboration? By Frank T. J. Bray and Michael Moodie. April 1977. 42pp. $5.00.

Iran's Quest for Security: U.S. Arms Transfers and the Nuclear Option. By Alvin J. Cottrell and James E. Dougherty. May 1977. 59pp. $5.00.

Ethiopia, the Horn of Africa, and U.S. Policy. By John H. Spencer. September 1977. 69pp. $5.00.

Beyond the Arab-Israeli Settlement: New Directions for U.S. Policy in the Middle East. By R. K. Ramazani. September 1977. 69pp. $5.00.

Spain, the Monarchy and the Atlantic Community. By David C. Jordan. June 1979. 55pp. $5.00.

Books

Atlantic Community in Crisis: A Redefinition of the Atlantic Relationship. Edited by Walter F. Hahn and Robert L. Pfaltzgraff, Jr. Pergamon Press, 1979. 386pp. $37.50.

Soviet Military Strategy in Europe. By Joseph D. Douglass, Jr. Pergamon Press, 1980. 252pp. $30.00.